Aram V. Arutyunov, Valeri Obukhovskii

Convex and Set-Valued Analysis

De Gruyter Graduate

Also of Interest

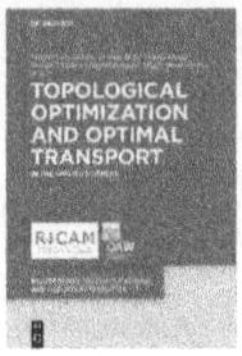

Topological Optimization and Optimal Transport.
In the Applied Sciences
M. Bergounioux, É. Oudet, M. Rumpf, G. Carlier, T. Champion, F. Santambrogio (Eds.), 2016
ISBN 978-3-11-043926-7, e-ISBN (PDF) 978-3-11-043041-7,
e-ISBN (EPUB) 978-3-11-043050-9

Variational Methods. In Imaging and Geometric Control
M. Bergounioux, G. Peyré, C. Schnörr, J.-B. Caillau, T. Haberkorn (Eds.), 2016
ISBN 978-3-11-043923-6, e-ISBN (PDF) 978-3-11-043039-4,
e-ISBN (EPUB) 978-3-11-043049-3

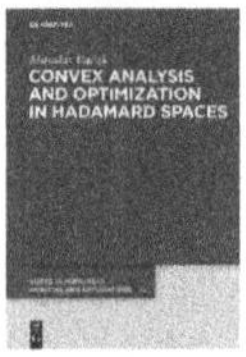

Convex analysis and optimization in Hadamard spaces
Miroslav Bacak, 2014
ISBN 978-3-11-036103-2, e-ISBN (PDF) 978-3-11-036162-9,
e-ISBN (EPUB) 978-3-11-039108-4

De Gruyter Series in Nonlinear Analysis and Applications
Jürgen Appell et al. (Eds.)
ISSN 0941-813X

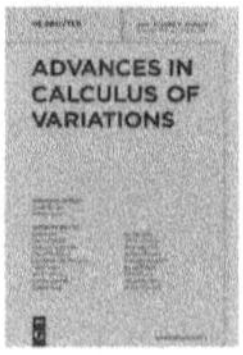

Advances in Calculus of Variations
Frank Duzaar, Juha Kinnunen (Managing Editors)
ISSN 1864-8258, eISSN 1864-8266

Aram V. Arutyunov, Valeri Obukhovskii

Convex and Set-Valued Analysis

Selected Topics

DE GRUYTER

Mathematics Subject Classification 2010
Primary: 47N10, 49J52, 49J53; Secondary: 28B20, 34A60, 47H04, 47H09, 47H10, 47H11, 54C60, 54C65, 54H25, 58C07, 90C05

Authors
Prof. Dr. Aram V. Arutyunov
RUDN University
117198 Moscow
Russian Federation
arutun@orc.ru

Prof. Dr. Valeri Obukhovskii
Voronezh State Pedagogical University
394043 Voronezh
Russian Federation
valerio-ob2000@mail.ru

ISBN 978-3-11-046028-5
e-ISBN (PDF) 978-3-11-046030-8
e-ISBN (EPUB) 978-3-11-046041-4

Library of Congress Cataloging-in-Publication Data
A CIP catalog record for this book has been applied for at the Library of Congress.

Bibliographic information published by the Deutsche Nationalbibliothek
The Deutsche Nationalbibliothek lists this publication in the Deutsche Nationalbibliografie; detailed bibliographic data are available on the Internet at http://dnb.dnb.de.

Cover image: Professor Sergei Zhukovskii
Typesetting: PTP-Berlin, Protago-TEX-Production GmbH, Berlin
Printing and binding: CPI books GmbH, Leck
♾ Printed on acid-free paper
Printed in Germany

www.degruyter.com

Preface

More than twenty years ago, the head of the Chair of System Analysis of the Faculty of Computational Mathematics and Cybernetics of the Moscow State University, academician A. B. Kurzhanskii suggested that the first author of this book should develop the course "Convex and set-valued analysis". As the result of the systematization of the material of these lectures, which the first author has been delivering to third year students at the MSU since 1993, the textbook [1] has appeared. This Russian edition constitutes the basis of the present monograph, which may be considered as its extension and addition. The present book contains several new important subjects. In particular, some of them are based on certain parts of the course "Introduction to set-valued analysis" which the second author is lecturing for fourth year students in the Faculty of Mathematics of the Voronezh State University.

The present book consists of two interconnected parts. The first part is devoted to convex analysis. This term was formed in the second half of the last century and it means a young branch of mathematics which is closely related to analysis and geometry. The formation of convex analysis as an independent branch was due to the rapid development of optimization theory, mathematical economics and the various areas of applied mathematics, in all of which an increasingly important role is played by the methods associated with the notion of convexity.

The subject of the second part is set-valued analysis. This is also a comparatively young branch of mathematics deeply connected with functional analysis and topology and intensively developing thanks to interesting applications, especially, in control theory, theory of optimization, theory of games, mathematical economics and other branches of contemporary mathematics. The main subject of set-valued analysis is the investigation of maps, which assign to each point of the original set not a single point of the image space, but the whole of its subset. Such maps are called set-valued (or multi-valued).

A large part of the presented material is quite traditional and is based on approaches partially developed in a number of books, some of which will be listed here. Concerning convex analysis, first of all, we would like to mention the monograph of R. T. Rockafellar [2] which since its publication in 1970 became the "encyclopedia of finite-dimensional convex analysis". Then appeared the books of V. M. Alekseev, V. M. Tikhomirov and S. V. Fomin [3], J. M. Borwein and A. S. Lewis [4], C. Castaing and M. Valadier [5], I. Ekeland and R. Temam [6], A. D. Ioffe and V. M. Tikhomirov [7], G. G. Magaril-Il'yaev and V. M. Tikhomirov [8], B. S. Mordukhovich [9], B. S. Mordukhovich and N. M. Nam [10], W. P. Odyniec and W. A. Slezak [11], B. N. Pshenichnyi [12], E. S. Polovinkin and M. V. Balashov [13], R. T. Rockafellar and R. J.-B. Wets [14], and others.

DOI 10.1515/9783110460308-001

Among the books on set-valued analysis we would like to distinguish the following works: J.-P. Aubin and A. Cellina [15], J.-P. Aubin and H. Frankowska [16], Yu. G. Borisovich, B. D. Gelman, A. D. Myshkis, and V. V. Obukhovskii [17], J. M. Borwein and A. S. Lewis [4], K. Deimling [18, 19], L. Górniewicz [20], S. Hu and N. S. Papageorgiou [21], M. Kamenskii, V. Obukhovskii, and P. Zecca [22], E. S. Polovinkin [23], J. Warga [24], and others.

At the same time, the present book does not repeat the above listed publications, but complements them. Thus it contains a number of results that are slightly touched on (or not covered at all) in the existing educational literature, (e.g., the theory of coincidences, properties of locally convex sets, etc.). A distinctive feature of this book is the large number of examples and exercises, clarifying the concepts introduced and the results obtained. Based on the courses of lectures, the book is maximally adapted to contemporary requirements and students' interests. The exposition is sufficiently closed in itself and does not require severe pretreatment (perhaps with the exception of one or two sections in the second part). The main information used from functional analysis is available in the classical sources (see, e.g., [25, 26, 27, 28, 29, 30, 31]).

The book is directed to a wide range of students educated in mathematics and applied mathematics as well as to post-graduate students and researchers. It is not written with a narrow purpose and is not aimed at any special group of readers (e.g., engineers or economists). It is designed for a wider audience, since it is very expediently and useful to possess the foundations of convex and set-valued analysis not only for mathematicians of various classical directions but also for experts in applied areas, especially in optimization problems.

The authors are grateful to their colleagues Boris Gelman, Alexey Izmailov, Boris Pasynkov and Evgenii Polovinkin for valuable fruitful discussions. We are obliged to Sergey Zhukovsky, Dmitrii Karamzin, Zukhra Zhukovskaya who attentively read the manuscript, helped to correct a number of inaccuracies and misprints and to prepare the manuscript for publication. We are also very grateful to the listeners of our lectures for their attention and enthusiasm.

We are obliged to our wives for their understanding and support. Last, we have the pleasure to express our thanks to the editors of the Walter de Gruyter publishing house for their constructive cooperation.

Moscow and Voronezh, November 2016

Aram V. Arutyunov
Valeri Obukhovskii

Contents

Part I: **Convex analysis**

1 Convex sets and their properties

First, for the reader's convenience and the completeness of the presentation, let us recall the definitions of certain standard objects, such as real linear space, normed, Euclidean space and so on. But beforehand, let us make an important remark. Namely, without loss of understanding, the reader may suppose, for simplicity, that the spaces considered in most of this chapter are finite-dimensional arithmetical spaces.

Recall that *a real linear space* X is the commutative group with respect to addition and for each element (called *vector*) $x \in X$ and a real number $\alpha \in \mathbb{R}$ the multiplication $\alpha x \in X$ is defined, satisfying the following axioms: for arbitrary $\alpha, \beta \in \mathbb{R}$, $x, y \in X$ we have

- *distributivity with respect to the addition of vectors*: $\alpha(x + y) = \alpha x + \alpha y$;
- *distributivity with respect to the addition of scalars*: $(\alpha + \beta)x = \alpha x + \beta x$;
- *associativity*: $(\alpha\beta)x = \alpha(\beta x)$;
- *the property of unity*: $1 \cdot x = x$.

We may consider the following examples of real linear spaces.

(1) $C[a, b]$, the space of continuous functions on the interval $[a, b]$.

(2) $\mathbb{R}^n$, the set of all ordered collections of n real numbers (or n-dimensional arithmetical space).

(3) The space of rectangular $n \times m$ matrices.

(4) $C^k[a, b]$, the space of functions continuous on the interval $[a, b]$ and k times continuously differentiable on the interval (a, b).

A real linear space X is called *normed* if it is endowed with the norm $\|\cdot\|$ satisfying usual axioms:

(i) $\|x\| \geq 0 \; \forall x, \quad \|x\| = 0 \Leftrightarrow x = 0$,

(ii) $\|\alpha x\| = |\alpha| \|x\| \quad \forall \alpha \in \mathbb{R}$,

(iii) $\|x + y\| \leq \|x\| + \|y\| \quad \forall x, y \in X$.

A sequence of points $\{x_i\}$ in a normed space *converges to a point* x_0 if $\|x_i - x_0\| \to 0$, $i \to \infty$. A sequence $\{x_i\}$ is called *fundamental* provided for each $\varepsilon > 0$ there exists such an integer N that $\|x_n - x_m\| < \varepsilon$ for all $n, m > N$. A normed space is called *complete* if every fundamental sequence of its points converges to a point of that space.

A complete normed space is called *Banach*[1] *space*. A real linear space in which the scalar product is introduced is called *Euclidean*[2]. In a Euclidean space the norm is defined by the formula $\|x\| = |x|$, where $|x|$ is the modulus of the vector x, which is given by the formula $|x| = (\langle x, x\rangle)^{1/2}$, and the angle brackets denote the scalar product.

1 Stefan Banach (1892–1945), a Polish mathematician.
2 Euclid, about 300 BC, an ancient Greek mathematician.

DOI 10.1515/9783110460308-002

A Euclidean space which is complete with respect to the above norm is called a *Hilbert*[3] *space*. In the following, in Euclidean spaces and, in particular, in $\mathbb{R}^n$ we will use, for convenience, the modulus of a vector $|x|$ instead of its norm.

The classical example of an infinite-dimensional Hilbert space presents the space l_2 consisting of all sequences of real numbers $x = (\xi_1, \xi_2, \dots)$ for which the series $\sum_{i=1}^{\infty}(\xi_i)^2$ converges. For $x = (\xi_1, \xi_2, \dots)$ and $y = (\eta_1, \eta_2, \dots)$ in l_2 *the scalar product* is defined by the formula $\langle x, y\rangle = \sum_{i=1}^{\infty} \xi_i, \eta_i$ and *the modulus of the vector x (its norm)* is given as $|x| = (\sum_{i=1}^{\infty}(\xi_i)^2)^{1/2}$.

Let X be a normed space. For $\varepsilon > 0$ and $x_0 \in X$ by $O(x_0, \varepsilon) = \{x \in X \colon \|x - x_0\| < \varepsilon\}$ we will denote *the open ε-neighborhood of a point* x_0. Let A be a subset of X. A point x_0 is called *an interior point of the set A* if there exists $\varepsilon > 0$ such that $O(x_0, \varepsilon) \subset A$. The set of all internal points of a set A will be denoted as $intA$ and called *the interior of A*. We will use also the notation $O_\varepsilon = O(0, \varepsilon)$.

A set A is called *open* if all its points are interior, i.e., $A = \operatorname{int} A$.

A set $C \subset X$ is called *closed* if it contains all its limit points, i.e., if a sequence $\{x_i\}$ lies in the set C and converges to a point x_0, then $x_0 \in C$. *The closure of a set C* is formed by the addition to the set C of all its limit points. It is easy to see that the closure of a set C coincides with the intersection of all closed sets containing C. The closure of a set C is denoted as $\operatorname{cl} C$. One may verify that a set C is closed if and only if its complement $A = X \setminus C$ is open.

A system of open sets $\{U_j\}_{j\in J}$ in X is called *an open covering of a set* $K \subset X$ provided $K \subset \bigcup_{j\in J} U_j$. A set K is called *compact* if each of its open coverings contains a finite subcovering. If a set is compact then it is closed and bounded, but the converse is not true, in general. Nevertheless, if $X = \mathbb{R}^n$, then each closed bounded set $A \subset X$ is compact.

For two given points x_1, x_2, the set of points x represented in the form $x = \alpha x_1 + (1 - \alpha)x_2$, where $\alpha \in [0, 1]$, is called *the interval joining points* x_1 and x_2, and is denoted as $[x_1, x_2]$. By $[x_1, x_2)$ we denote a semi-interval consisting of points x represented in the form $x = \alpha x_1 + (1 - \alpha)x_2$, where $0 < \alpha \le 1$. The semi-interval $(x_1, x_2]$ and interval (x_1, x_2) can be defined similarly.

Now, let X be a real linear space and $A \subseteq X$.

Definition 1.1. A set A is called convex if for each two points $x_1, x_2 \in A$ we have

$$[x_1, x_2] \subset A \quad (\text{i.e., } \alpha x_1 + (1 - \alpha)x_2 \in A \quad \forall \alpha \in [0, 1]).$$

An empty set is supposed to be convex by definition.

From the above definition it follows that each singleton $\{a\}$ is a convex set.

3 David Hilbert (1862–1943), a German mathematician.

Definition 1.2. The sum $\sum_{i=1}^{k} \alpha_i x_i$ is called the convex combination of points $x_1, \ldots, x_k$ if $\alpha_i \geq 0$, $i = 1, \ldots, k$, $\sum_{i=1}^{k} \alpha_i = 1$.

Let $A, B \subseteq X$ and $\alpha \in \mathbb{R}$. Consider the sets

$$A + B = \{x \in X \colon x = a + b,\ a \in A,\ b \in B\},$$
$$\alpha A = \{x \in X \colon x = \alpha a,\ a \in A\}.$$

The set $A + B$ is called *the sum (in the sense of Minkowski*[4]*) of the sets A and B*, and the set αA is called *the product of the number α by the set A*. It is evident that $O(x_0, \varepsilon) = x_0 + O_\varepsilon$.

Proposition 1.3. *The following properties hold true:*

(1) *The intersection of any collection of convex sets $A_j \subseteq X$, $j \in J$ is the convex set.*

(2) *Let $A_1, \ldots, A_n$ be convex subsets of X, $\alpha_1, \ldots, \alpha_n \in \mathbb{R}$. Then the set $\sum_{i=1}^{n} \alpha_i A_i$ is convex.*

Proof. (1) Take arbitrary points $x, y \in \bigcap_{j \in J} A_j$. Each of the sets A_j is convex. So, $[x, y] \subset A_j$ for each $j \in J$. Hence $[x, y] \subset \bigcap_{j \in J} A_j$, that means, by definition, the convexity of the intersection of sets $\bigcap_{j \in J} A_j$.

(2) Take arbitrary points $x, y \in \sum_{i=1}^{n} \alpha_i A_i$. By definition, there exist $x_1, y_1 \in A_1, \ldots, x_n, y_n \in A_n$ such that

$$x = \sum_{i=1}^{n} \alpha_i x_i, \quad y = \sum_{i=1}^{n} \alpha_i y_i.$$

From the convexity of the sets A_i it follows that for each $\alpha, \beta \geq 0 \colon \alpha + \beta = 1$ we have $\alpha x_i + \beta y_i \in A_i$ and hence, $\alpha x + \beta y = \sum_{i=1}^{n} \alpha_i(\alpha x_i + \beta y_i) \in \sum_{i=1}^{n} \alpha_i A_i$, implying the convexity of the set $\sum_{i=1}^{n} \alpha_i A_i$. ☐

The definitions of operations of sum and product of sets immediately imply the inclusion

$$(\alpha + \beta)A \subseteq \alpha A + \beta A$$

which is true for an arbitrary set A and numbers α, β.

Proposition 1.4. *Let A be a convex set. Then for each $\alpha \geq 0$, $\beta \geq 0$ the following formula holds true*

$$(\alpha + \beta)A = \alpha A + \beta A. \tag{1.1}$$

Proof. From the above mentioned it follows that it is sufficient to prove the inclusion $\alpha A + \beta A \subseteq (\alpha + \beta)A$. If $\alpha + \beta = 0$, then $\alpha = \beta = 0$ and, hence, this inclusion is evidently true. Consider the case $\alpha + \beta > 0$. Let $\xi \in \alpha A + \beta A$. Then $\xi = \alpha x_1 + \beta x_2$ for certain

4 Hermann Minkowski (1864–1909), a German mathematician.

$x_1, x_2 \in A$, and, by virtue of convexity of A, we have

$$\xi = (\alpha + \beta)\left(\frac{\alpha}{(\alpha + \beta)}x_1 + \frac{\beta}{(\alpha + \beta)}x_2\right) \in (\alpha + \beta)A,$$

and the proof of the inclusion is complete. □

If the set A is not convex, then equality (1.1) may be violated. In fact, let A be the unit circle and $\alpha = \beta = 1$. Then $2A$ is the circle of radius 2 with the center at the origin and $A + A$ is the disk of the radius 2 with the center at the origin. Similarly, if numbers α and β have different signs, then equality (1.1) can also be violated: for example take A a unit disk and $\alpha = 1, \beta = -1$.

Proposition 1.5. *A convex set A contains each convex combination of its points.*

Proof. It is necessary to show that for every $n \geq 2$ the relations

$$x = \sum_{i=1}^{n} \alpha_i x_i, \quad x_i \in A, \ \alpha_i \geq 0, \ \sum_{i=1}^{n} \alpha_i = 1,$$

imply $x \in A$.

The proof will be given by induction. For $n = 2$ the assertion follows from the definition of a convex set. Let the statement be true for $n = r$. Let us verify it for $n = r + 1$.

Without loss of generality we will suppose that $\sum_{i=1}^{r} \alpha_i > 0$. Then

$$\sum_{i=1}^{r+1} \alpha_i x_i = \sum_{i=1}^{r} \alpha_i \left(\sum_{i=1}^{r} \frac{\alpha_i}{\sum_{i=1}^{r} \alpha_i} x_i \right) + \alpha_{r+1} x_{r+1} = \hat{\alpha}\hat{x} + \alpha_{r+1} x_{r+1} \in A$$

by virtue of convexity of A. Here

$$\hat{x} = \sum_{i=1}^{r} \frac{\alpha_i}{\sum_{i=1}^{r} \alpha_i} x_i \in A$$

by virtue of convexity of A and the induction assumption and $\hat{\alpha} = \sum_{i=1}^{r} \alpha_i$, implying $\hat{\alpha} + \alpha_{r+1} = 1$. □

2 The convex hull of a set. The interior of convex sets

The convex hull of a set

Let, as earlier, X be a real linear space and $A \subseteq X$.

Definition 2.1. A convex hull of a set A is the intersection of all convex sets containing A. It is denoted by $\operatorname{conv} A$.

From the definition it immediately follows that the convex hull $\operatorname{conv} A$ is the least (with respect to the inclusion) convex set containing the set A.

Theorem 2.2. *The set* $\operatorname{conv} A$ *consists of those and only those points which may be represented as convex combinations of a finite number of points from A.*

Proof. Denote by B the set of all possible convex combinations of a finite number of points from A. By virtue of Proposition 1.5 we have $B \subset \operatorname{conv} A$, since the set $\operatorname{conv} A$ is convex and $A \subseteq \operatorname{conv} A$. Let us prove the inverse inclusion.

We show that the set B is convex. In fact, let $b_1, b_2 \in B$. Then each of these points can be represented in the form of a convex combination of a finite number of points from A. Moreover, increasing, if necessary, the number of these points, we can assume, without loss of generality, that there exist a number m and points $a_i \in A$, $i = \overline{1, m}$, for which we have the representations $b_s = \sum_{i=1}^{m} \alpha_{s,i} a_i$, $s = 1, 2$. Here $\alpha_{s,i}$, $i = \overline{1, m}$, $s = 1, 2$, are certain nonnegative numbers for which $\sum_{i=1}^{m} \alpha_{s,i} = 1$, $s = 1, 2$. We should show that $\theta b_1 + (1 - \theta) b_2 \in B$, $\forall \theta \in [0, 1]$. In fact,

$$\theta b_1 + (1 - \theta) b_2 = \sum_{i=1}^{m} \left(\theta \alpha_{1,i} + (1 - \theta)\alpha_{2,i}\right) a_i \in B,$$

since $(\theta \alpha_{1,i} + (1 - \theta)\alpha_{2,i}) \geq 0$ for all i and

$$\sum_{i=1}^{m} (\theta \alpha_{1,i} + (1 - \theta)\alpha_{2,i}) = \theta \sum_{i=1}^{m} \alpha_{1,i} + (1 - \theta) \sum_{i=1}^{m} \alpha_{2,i} = \theta + (1 - \theta) = 1.$$

So, the convexity of the set B is proved. At the same time, it is evident that $A \subset B$, and hence $\operatorname{conv} A \subseteq B$. So, the equality $\operatorname{conv} A = B$ is true. □

In the case when $X = \mathbb{R}^n$, the above assertion can be essentially refined.

Theorem 2.3 (Carathéodory[1]). *Let $A \subset \mathbb{R}^n$. Then* $\operatorname{conv} A$ *consists of all possible convex combinations consisting of no more than $n + 1$ points of the set A.*

1 Constantin Carathéodory (1873–1950), a German mathematician.

DOI 10.1515/9783110460308-003

Proof. Set

$$B = \left\{ x \colon x = \sum_{i=1}^{n+1} \alpha_i x_i, \text{ where } x_i \in A,\ \alpha_i \geq 0\ \forall i = \overline{1, n+1},\ \sum_{i=1}^{n+1} \alpha_i = 1 \right\}.$$

By Theorem 2.2 we have $B \subseteq \operatorname{conv} A$. It remains to show that $\operatorname{conv} A \subseteq B$.

Let $x \in \operatorname{conv} A$. Then, by Theorem 2.2, for some natural number r we have the representation

$$x = \sum_{i=1}^{r+1} \alpha_i x_i, \quad \text{where } \alpha_i \geq 0,\ x_i \in A,\ i = \overline{1, r+1},\ \sum_{i=1}^{r+1} \alpha_i = 1.$$

If $r \leq n$, then $x \in B$. Now, let $r > n$. We show that in this case x may be represented as a convex combination of no more than r points of the set A. If at least one of the numbers α_i equals zero, it is evident. Now, let $\alpha_i > 0\ \forall i$. Since $r > n$, the system of vectors $(x_i - x_{r+1})$, $i = 1, \ldots, r$ is linearly dependent. Whence, there exist such numbers $t_1, \ldots, t_r$, not all equal to zero, that $\sum_{i=1}^{r} t_i(x_i - x_{r+1}) = 0$.

Set $t_{r+1} = -\sum_{i=1}^{r} t_i$. Then $\sum_{i=1}^{r+1} x_i t_i = 0$, $\sum_{i=1}^{r+1} t_i = 0$ and, hence, for an arbitrary number c we have

$$x = \sum_{i=1}^{r+1} \alpha_i x_i + \sum_{i=1}^{r+1} c\, t_i x_i = \sum_{i=1}^{r+1} (\alpha_i + c\, t_i) x_i, \quad \sum_{i=1}^{r+1} (\alpha_i + c\, t_i) = 1. \tag{2.1}$$

It is evident that at least one of the numbers t_i is negative. Moreover, $\alpha_i > 0$ for each i and, hence, for $c = 0$, all numbers $(\alpha_i + c\, t_i)$ are positive. Let us increase the parameter c from zero to infinity. It is clear then that there exists the least number $c > 0$ such that for all numbers i we have $(\alpha_i + c\, t_i) \geq 0$, and for a certain $i_0 \leq r + 1$ we have $(\alpha_{i_0} + c\, t_{i_0}) = 0$. So, throwing off the i_0-th summand $(\alpha_{i_0} + c\, t_{i_0})x_{i_0}$ in the representation (2.1), we get the desired assertion, namely, that x can be represented in the form of a convex combination of no more than r points of the set A.

Repeating this procedure a finite number of times, we obtain that $r \leq n$ and, hence, $x \in B$. So, we proved that $\operatorname{conv} A \subseteq B$. □

Theorem 2.4. *Let $A \subset \mathbb{R}^n$ be a compact set. Then* conv *A is also a compact set.*

Proof. Take an arbitrary sequence $\{x_k\} \subset \operatorname{conv} A$. Then, by the Carathéodory theorem, the following representation holds

$$x_k = \sum_{i=1}^{n+1} \alpha_k^i y_k^i, \quad \text{where } \alpha_k^i \geq 0,\ \sum_{i=1}^{n+1} \alpha_k^i = 1,\ y_k^i \in A\ \forall i, k.$$

Since $0 \leq \alpha_k^i \leq 1$, for each $i = 1, \ldots, n+1$ from the sequence $\{\alpha_k^i\}$ we may choose a subsequence converging to a certain point of the interval $[0, 1]$. Similarly, by virtue of compactness of A, for each number i from the sequence $\{y_k^i\}$ we may choose a subsequence converging to a certain point of the set A.

Choosing, step by step, such subsequences and denoting them by $\{\alpha^i_k\}$, $\{y^i_k\}$, we obtain that under $k \to \infty$, for each number i we have $\alpha^i_k \to \alpha^i \in [0, 1]$, $y^i_k \to y^i \in A$. Then $x_k \to x$, $k \to \infty$, where

$$x = \sum_{i=1}^{n+1} \alpha^i y^i, \quad y^i \in A, \; \alpha_i \geq 0 \; \forall i, \; \sum_{i=1}^{n+1} \alpha^i = 1.$$

Whence, $x \in \operatorname{conv} A$. Therefore, it is proved that from each sequence $\{x_k\}$, belonging to $\operatorname{conv} A$ a subsequence converging to a certain point of the set $\operatorname{conv} A$ may be chosen. It means that the set $\operatorname{conv} A$ is compact. □

Exercise 2.5.

(1) *Is the convex hull of a closed subset of $\mathbb{R}^n$ closed?*

(2) *Is the convex hull of a bounded set bounded?*

Decision.

(1) *No, it is not. As an example, we can consider in the space $\mathbb{R}^2$ the straight line given by the equation $y = 0$ (in the Cartesian system of coordinates Oxy) and the point $(0, 1)$. This line and the point form a closed set A, but at the same time the set*

$$\operatorname{conv} A = \{(x, y) : x \in \mathbb{R}, \; y \in [0, 1)\} \cup \{(0, 1)\}$$

is not closed.

(2) *Yes, it is. In fact, if the set A is bounded, then there exists a ball B centered at the origin containing the set A. The ball is a convex set, hence $\operatorname{conv} A \subset B$. Therefore, the convex hull is a bounded set.*

Let us return to Theorem 2.4. The assumption that the space X is finite-dimensional is essential. Namely, the Mazur[2] theorem (see, e.g., [26], Theorem V.2.6) claims that if X is a Banach space and $A \subset X$ is compact, then the set $\operatorname{cl}(\operatorname{conv} A)$ is also compact. At the same time, the following example demonstrates that, if a Banach space X is infinite-dimensional, then the convex hull of a compact set may be not a closed set, and hence it may be not compact.

Example 2.6. Let $X = l_2$. Consider the set $A \subset X$, consisting of zero and points x_i, $i = 1, 2, \ldots$, where x_i is the sequence in which the i-th member is equal to 2^{-i}, and the other members are zeros. It is easy to see that $x_i \to 0$ and hence A is compact.

By virtue of Theorem 2.2, if $x \in \operatorname{conv} A$, then the sequence $x = (x^1, x^2, \ldots)$ is finite (i.e., it has at least a finite number of members that differ from zero). Set $a_n = \gamma_n \xi_n$, where $\gamma_n = (\sum_{i=1}^n 2^{-i})^{-1}$, and ξ_n is the sequence whose i-th member is equal to 4^{-i}, if $i \leq n$, and zero, if $i > n$. It is easy to see that $a_n \in \operatorname{conv} A$, $\forall n$. It is evident that $\gamma_n \to 1$ and $\xi_n \to \bar{x}$ while $n \to \infty$, where $\bar{x}$ is the sequence whose i-th member is equal to 4^{-i}. Hence, $a_n \to \bar{x}$, $n \to \infty$. But the sequence $\bar{x}$ is not finite and, by the above mentioned, $\bar{x} \notin \operatorname{conv} A$, and hence the set $\operatorname{conv} A$ is not closed.

2 Stanislaw Mazur (1905–1981), a Polish mathematician.

The interior of convex sets

Let X be a normed space.

Theorem 2.7. *Let $A \subseteq X$ be a convex set, $x_1 \in \operatorname{int} A$ and $x_2 \in \operatorname{cl} A$. Then $[x_1, x_2) \subset \operatorname{int} A$.*

Proof. Consider an arbitrary point $x \in (x_1, x_2)$. Then

$$\exists \lambda \in (0,1) : x = (1-\lambda)x_1 + \lambda x_2 .$$

Since $x_1 \in \operatorname{int} A$, there exists $\varepsilon > 0$ such that $x_1 + O_\varepsilon \subset A$. Since $x_2 \in \operatorname{cl} A$, there exists a sequence $\{x_{2,n}\} \subset A$, such that $x_{2,n} \to x_2$, $n \to \infty$. By virtue of convexity of the set A we have

$$(1-\lambda)(x_1 + O_\varepsilon) + \lambda x_{2,n} = x_n + (1-\lambda)O_\varepsilon = x_n + O_{(1-\lambda)\varepsilon} \subset A,$$

where $x_n = (1-\lambda)x_1 + \lambda x_{2,n}$. It is clear that $x_n \to x$ as $n \to \infty$. Hence, for all sufficiently large n we have $\|x - x_n\| < (1-\lambda)\varepsilon/2$. From here it follows that $x + O_{(1-\lambda)\varepsilon/2} \subset A$ and, hence, $x \in \operatorname{int} A$. □

Corollary 2.8. *Let $A \subseteq X$ be a convex set. Then the set $\operatorname{int} A$ is convex. If, moreover, $\operatorname{int} A \neq \varnothing$, then*

$$\operatorname{cl}(\operatorname{int} A) = \operatorname{cl} A, \quad \operatorname{int} A = \operatorname{int}(\operatorname{cl} A).$$

Proof. In fact, since $\operatorname{int} A \subset \operatorname{cl} A$, then, by virtue of Theorem 2.7, if $x_1, x_2 \in \operatorname{int} A$, then

$$[x_1, x_2) \subset \operatorname{int} A \Rightarrow [x_1, x_2] \subset \operatorname{int} A,$$

implying the convexity of the set $\operatorname{int} A$.

We have $\operatorname{cl}(\operatorname{int} A) \subset \operatorname{cl} A$, since, obviously, $\operatorname{int} A \subset A$. Now, supposing $\operatorname{int} A \neq \varnothing$, prove the inverse inclusion.

Take an arbitrary $x \in \operatorname{cl} A$ and a certain $y \in \operatorname{int} A$, $y \neq x$. By Theorem 2.7, we have the inclusion $[y, x) \subset \operatorname{int} A$. This, in particular, implies that there exists the sequence $\{y_n\} \in [y, x) \subset \operatorname{int} A$,converging to x. This means that $x \in \operatorname{cl}(\operatorname{int} A)$. The inclusion $\operatorname{cl}(\operatorname{int} A) \supset \operatorname{cl} A$ is proved.

We have $\operatorname{int} A \subset \operatorname{int}(\operatorname{cl} A)$, since $A \subset \operatorname{cl} A$. Now, supposing $\operatorname{int} A \neq \varnothing$, prove the inverse inclusion.

Fix any $y \in \operatorname{int} A$. Take an arbitrary $x \in \operatorname{int}(\operatorname{cl} A)$. Then there exists $\varepsilon > 0$ such that $x + O_\varepsilon \subset \operatorname{cl} A$. By virtue of Theorem 2.7, for each $z \in x + O_\varepsilon$ the inclusion $[y, z) \subset \operatorname{int} A$ holds. Set $z = x - (y - x) \cdot \frac{\varepsilon}{2\|y-x\|}$. Then $z \in x + O_\varepsilon$. But $x \in (y, z)$, since

$$x = \lambda y + (1-\lambda)z, \quad \text{for } \lambda = \frac{\varepsilon}{\varepsilon + 2\|y - x\|} \quad \text{and } \lambda \in (0,1).$$

From $[y, z) \subset \operatorname{int} A$ it follows that $x \in \operatorname{int} A$. The inclusion $\operatorname{int} A \supset \operatorname{int}(\operatorname{cl} A)$ is proved. □

Theorem 2.9. *The convex hull of an open set $A \subseteq X$ is the open set.*

Proof. Since $A \subset \operatorname{conv} A$, we have $A = \operatorname{int} A \subset \operatorname{int}(\operatorname{conv} A)$, and, by Corollary 2.8, $\operatorname{int}(\operatorname{conv} A)$ is the convex set. Taking into account that $\operatorname{conv} A$ is the least convex set containing A, we have $\operatorname{conv} A \subset \operatorname{int}(\operatorname{conv} A)$. The inverse inclusion is evident. So, $\operatorname{conv} A$ is the open set. □

3 The affine hull of sets. The relative interior of convex sets

The affine hull

Recall that *the linear combination of points* $x_1, \dots, x_n \in X$ is any point x which may be represented in the form

$$x = \sum_{i=1}^{n} \alpha_i x_i, \quad \alpha_i \in \mathbb{R}.$$

The linear hull of a set $A \subseteq X$ is the set of all possible linear combinations of points from A, i.e., it has the form

$$\left\{ x\colon x = \sum_{i=1}^{n} \lambda_i x_i \;\middle|\; \lambda_i \in \mathbb{R},\ x_i \in A,\ n \in \mathbb{N} \right\}.$$

The linear hull of the set A will be denoted by $\operatorname{lin} A$. It is clear that $\operatorname{lin} A$ is the intersection of all linear subspaces containing A.

Let us give necessary definitions.

Definition 3.1. The affine combination of points $x_1, \dots, x_n \in X$ is a point which may be represented in the form

$$x = \sum_{i=1}^{n} \alpha_i x_i, \quad \text{where } \alpha_i \in \mathbb{R},\ \sum_{i=1}^{n} \alpha_i = 1.$$

It is obvious that each convex combination of points is their affine combination, but not vice versa. Each affine combination of points is their linear combination, but not vice versa.

Definition 3.2. The affine hull $\operatorname{aff} A$ of a set $A \subseteq X$ is the set of all possible affine combinations of points from A, i.e.,

$$\operatorname{aff} A = \left\{ x\colon x = \sum_{i=1}^{n} \alpha_i x_i \;\middle|\; n \in \mathbb{N},\quad \sum_{i=1}^{n} \alpha_i = 1,\ x_i \in A \right\}.$$

Definition 3.3. A set $A \subseteq X$ is called a linear manifold if for every points $x_1, x_2 \in A$, the set A contains the straight line passing through them:

$$\{x\colon x = \alpha x_1 + (1 - \alpha) x_2,\ \alpha \in \mathbb{R}\}.$$

By induction, it is easy to obtain that the set $A \subseteq X$ is a linear manifold if and only if

$$\sum_{i=1}^{n} \alpha_i x_i \in A \quad \forall n \in \mathbb{N},\ \forall x_i \in A,\ \forall \alpha_i \in \mathbb{R}\colon \sum_{i=1}^{n} \alpha_i = 1.$$

DOI 10.1515/9783110460308-004

It is also not complicated to verify that aff (aff A) = aff A, the set aff A is the linear manifold and, if A is the linear manifold, then aff $A = A$.

Theorem 3.4. *The following assertions hold true:*
(1) *The intersection of an arbitrary family of linear manifolds is the linear manifold.*
(2) *The linear manifold L is the linear subspace if and only if $0 \in L$.*

Proof. The first assertion is evident. Let us prove the second one. In fact, if L is the linear subspace, then $0 \in L$.

To the contrary, if L is the linear manifold and $0 \in L$, then for each $x_1, x_2 \in L$ and $\alpha \in \mathbb{R}$ we have

$$\alpha x_1 = \alpha x_1 + (1-\alpha) \cdot 0 \in L,$$
$$\frac{1}{2}x_1 + \frac{1}{2}x_2 \in L \Rightarrow x_1 + x_2 = 2\left(\frac{1}{2}x_1 + \frac{1}{2}x_2\right) \in L. \qquad \square$$

Theorem 3.5. *For an arbitrary set A, its affine hull* aff A *coincides with the intersection of all linear manifolds containing the set A.*

Proof. By B denote the intersection of linear manifolds containing the set A. It is clear that $B \subset$ aff A, since aff A is the linear manifold containing A. The set B itself is the linear manifold (as the intersection of linear manifolds) and $B \supset A$. Therefore $B =$ aff $B \supset$ aff A and, hence, $B \supset$ aff A. So, aff $A = B$. □

Theorem 3.6. *For an arbitrary point $x_0 \in A$ we have:*

$$\operatorname{aff} A = x_0 + \operatorname{lin}(A - x_0).$$

Proof. Let $x \in x_0 + \operatorname{lin}(A - x_0)$. Then there exist $n \in \mathbb{N}$, points $x_1, \ldots, x_n \in A$ and numbers $\alpha_1, \ldots, \alpha_n \in \mathbb{R}$ such that

$$x = x_0 + \sum_{i=1}^{n} \alpha_i (x_i - x_0).$$

Then $x = \sum_{i=0}^{n} \alpha_i x_i \in \operatorname{aff} A$, where $\alpha_0 = 1 - \sum_{i=1}^{n} \alpha_i$, and hence, $\sum_{i=0}^{n} \alpha_i = 1$. This means that $x_0 + \operatorname{lin}(A - x_0) \subseteq \operatorname{aff} A$.

Prove the inverse inclusion. Let $x \in \operatorname{aff} A$. Then there exist natural number n, points $x_1, \ldots, x_n \in A$ and numbers $\alpha_1, \ldots, \alpha_n \in \mathbb{R}$ such that

$$\sum_{i=1}^{n} \alpha_i = 1, \quad x = \sum_{i=1}^{n} \alpha_i x_i.$$

Therefore

$$x = x_0 + \sum_{i=1}^{n} \alpha_i (x_i - x_0) \in x_0 + \operatorname{lin}(A - x_0). \qquad \square$$

Corollary 3.7. *For arbitrary sets $A_1, A_2 \subseteq X$ the following equality holds*

$$\operatorname{aff}(A_1 + A_2) = \operatorname{aff} A_1 + \operatorname{aff} A_2.$$

Proof. For arbitrary sets B_1, B_2, such that $0 \in B_1$, $0 \in B_2$, it is evident that $B_1 \subseteq (B_1 + B_2)$, $B_2 \subseteq (B_1 + B_2)$, from which the equality $\operatorname{lin}(B_1 + B_2) = \operatorname{lin} B_1 + \operatorname{lin} B_2$ easily follows. Take arbitrary $x_1 \in A_1$, $x_2 \in A_2$. Then $\operatorname{aff} A_i = \operatorname{lin}(A_i - x_i) + x_i$, $i = 1, 2$. By virtue of the above mentioned, we have

$$\begin{aligned}\operatorname{aff} A_1 + \operatorname{aff} A_2 &= \operatorname{lin}(A_1 - x_1) + x_1 + \operatorname{lin}(A_2 - x_2) + x_2 \\ &= \operatorname{lin}((A_1 + A_2) - (x_1 + x_2)) + (x_1 + x_2) = \operatorname{aff}(A_1 + A_2).\end{aligned}$$ □

Let L be a linear manifold. It is easy to verify that, if $x_0 \in L$, then the set $L - x_0$ is the linear subspace and, moreover, $L - x_0 = L - x_1$ for every $x_0, x_1 \in L$. This justifies the following definition.

Definition 3.8. The dimension of a linear manifold L is the dimension of the linear subspace $(L - x_0)$, where x_0 is an arbitrary vector from L. The dimension of a convex set A is the dimension of the linear manifold $\operatorname{aff} A$. The dimension of a set A, as usual, will be denoted by $\dim A$.

Definition 3.9. A system of vectors $x_0, \dots, x_n \in X$ is called affinely independent if from the relations

$$\sum_{i=0}^{n} \alpha_i x_i = 0, \quad \sum_{i=0}^{n} \alpha_i = 0$$

it follows that $\alpha_i = 0 \ \forall i$.

Theorem 3.10. *A system of vectors $x_0, \dots, x_n \in X$ is affinely independent if and only if for each given number i_0, the vectors*

$$(x_i - x_{i_0}), \quad i = 0, \dots, n, \ i \neq i_0$$

are linearly independent.

Proof. Necessity. For determinacy, take $i_0 = 0$. Suppose that the vectors $(x_i - x_0)$, $i = 1, \dots, n$ are linearly dependent. Then there exist numbers $\alpha_1, \dots, \alpha_n$ not all equal to zero and such that

$$\sum_{i=1}^{n} \alpha_i (x_i - x_0) = 0.$$

Set $\alpha_0 = -\sum_{i=1}^{n} \alpha_i$. Then $\sum_{i=0}^{n} \alpha_i x_i = 0$, $\sum_{i=0}^{n} \alpha_i = 0$, where not all coefficients α_i are equal to zero. It means that the system of vectors $x_0, \dots, x_n$ is affinely dependent. We obtain the contradiction.

The proof of *sufficiency* is similar. □

From the proved theorem it follows that in the n-dimensional space $\mathbb{R}^n$ the maximal number of affine independent vectors is equal to $(n + 1)$.

Now, let in a system of vectors $x_i \in \mathbb{R}^n$, $i = \overline{0, k}$ be given, where $k \leq n$. Then the affine independence of this system is the generic position condition. It means that, firstly, if this system is affinely independent, then each system consisting of sufficiently close vectors $x_{i,\varepsilon}$, $i = \overline{0,k}$ (i.e., there exists $\varepsilon > 0$ such that $|x_{i,\varepsilon} - x_i| \le \varepsilon\ \forall i$) is also affinely independent. Secondly, if the initial system is affinely dependent, then there exists a system, arbitrarily close to it, which is affinely independent. The validity of this assertion follows from Theorem 3.10 which reduces the property of affine independence to the linear independence of vectors and to the fact known from linear algebra that for a system consisting of no more than n vectors from $\mathbb{R}^n$, linear independence is the generic position property.

Proposition 3.11. *If a system of vectors x_i, $i = \overline{0,n}$, is affinely independent, $x = \sum_{i=0}^n \alpha_i x_i$ and $\sum_{i=0}^n \alpha_i = 1$, then the numbers α_i are uniquely defined.*

Proof. In fact, let the following representations hold true

$$x = \sum_{i=0}^n \alpha_i x_i, \ \sum_{i=0}^n \alpha_i = 1, \quad x = \sum_{i=0}^n \beta_i x_i, \ \sum_{i=0}^n \beta_i = 1.$$

Then $0 = \sum_{i=0}^n \gamma_i x_i$, $\sum_{i=0}^n \gamma_i = 0$, where $\gamma_i = \alpha_i - \beta_i$. By virtue of the affine independence of the system x_i, $i = \overline{0,n}$ it follows that $\alpha_i = \beta_i\ \forall i$. □

Let a system of vectors $x_i \in \mathbb{R}^n$, $i = \overline{0,n}$ be affinely independent. Then each vector $x \in \mathbb{R}^n$ may be represented in the form $x = \sum_{i=0}^n \alpha^i x_i$, $\sum_{i=0}^n \alpha^i = 1$, and, moreover, by virtue of Proposition 3.11, this representation is unique. Numbers $\alpha^0, \dots, \alpha^n$ are called *barycentric coordinates of a point* x with respect to the system $x_0, \dots, x_n$.

Theorem 3.12. *Let L be the linear manifold*

$$L = \left\{ \alpha = (\alpha^0, \dots, \alpha^n) \in \mathbb{R}^{n+1} : \sum_{i=0}^n \alpha^i = 1 \right\}$$

and the map $\alpha\colon \mathbb{R}^n \to L$ be defined as $\alpha(x) = (\alpha^0, \dots, \alpha^n)$, the barycentric coordinates of a point $x \in \mathbb{R}^n$ with respect to an affinely independent system $x_0, \dots, x_n$. Then the map $\alpha(\cdot)$ is continuous.

Proof. Suppose that a sequence of points $\{\xi_j\}$, converges to a certain point x. We will show that $\alpha(\xi_j) \to \alpha(x)$ as $j \to \infty$.

Set $\alpha_j = \alpha(\xi_j)$. First, prove that the sequence $\{\alpha_j\}$ is bounded. In fact, supposing the contrary, we will have, passing to a subsequence, that $|\alpha_j| \to \infty$ as $j \to \infty$. We obtain

$$\begin{aligned} |\xi_j| = |\alpha_j^0 x_0 + \dots + \alpha_j^n x_n| &= |\alpha_j^0(x_0 - x_n) + \dots + \alpha_j^{n-1}(x_{n-1} - x_n) + x_n| \\ &\ge |\alpha_j^0(x_0 - x_n) + \dots + \alpha_j^{n-1}(x_{n-1} - x_n)| - |x_n|, \end{aligned}$$

where the coordinates of the vector α_j are denoted by upper symbols.

The converging sequence $\{\xi_j\}$ is bounded, i.e., for a certain m, the inequality $|\xi_j| \le m$ holds for all j. Define the linear operator A by the formula $A\alpha = \sum_{i=0}^{n-1} \alpha^i(x_i - x_n)$. Then $|A\alpha_j| \le |x_n| + m = M$.

Set $e_j = \alpha_j/|\alpha_j|$. Then $|Ae_j| \leq M/|\alpha_j| \to 0$. Passing in the sequence $\{e_j\}$ to a subsequence, we will assume that $e_j \to e$. Tending $j \to \infty$, we get $Ae = 0$, $e \neq 0$, contradicting, by virtue of Theorem 3.10 on the affine independence of the system x_i, $i = \overline{0, n}$. So, we proved that the sequence $\{\alpha_j\}$ is bounded.

Now, let $\{\alpha_s\}$ be any subsequence of the sequence $\{\alpha_j\}$, and $\alpha_s \to \hat{\alpha}$. For $\alpha = (\alpha^0, \dots, \alpha^n) \in L$, set $\phi(\alpha) = \sum_{i=0}^n \alpha^i x_i$. It is clear that the map $\phi(\cdot)$ is the inverse to the map $\alpha(\cdot)$. Then, since $\alpha_s = \alpha(\xi_s)$, we have $\phi(\alpha_s) = \phi(\alpha(\xi_s)) = \xi_s$ $\forall s$. By continuity of the map ϕ we obtain $\phi(\alpha_s) \to \phi(\hat{\alpha})$, from which, $\xi_s \to x$ implies $\phi(\hat{\alpha}) = x \Rightarrow \hat{\alpha} = \alpha(x)$. We see that the initial bounded sequence $\{\alpha_j\}$ has only one limit point $\alpha(x)$, and, therefore, the whole this sequence converges to this point. □

Definition 3.13. Let vectors $x_0, \dots, x_n \in X$ be affinely independent. The set $S = \operatorname{conv}\{x_0, \dots, x_n\}$ is called the n-dimensional simplex with vertices $x_0, \dots, x_n$.

Theorem 3.14. *Each n-dimensional simplex in the space $\mathbb{R}^n$ has a nonempty interior.*

Proof. Let $S = \operatorname{conv}(x_0, \dots, x_n)$. Passing from the simplex S to the simplex $S - (\sum_{i=0}^n x_i)(n+1)^{-1}$, we will assume, without loss of generality, that $\sum_{i=0}^n x_i = 0$.

Prove the existence of such $\varepsilon_0 > 0$, that $\varepsilon x \in S$ for each $\varepsilon \in (0, \varepsilon_0)$ and $x \in O_1 = O(0, 1)$. Denote by $(\lambda^0(x), \dots, \lambda^n(x))$ the barycentric coordinates of the point x with respect to the system $\{x_0, \dots, x_n\}$. By virtue of Theorem 3.12, the barycentric coordinates of all points $x \in O_1$ are bounded in modulo. So, there exists $\varepsilon_0 > 0$ such that $\frac{1-\varepsilon}{n+1} + \varepsilon\lambda^i(x) \geq 0$ $\forall i$ for all $x \in O_1$ and all $\varepsilon \in (0, \varepsilon_0)$. We have

$$\begin{aligned}
\varepsilon x &= 0\frac{1-\varepsilon}{n+1} + \varepsilon x = \sum_{i=0}^{n}\left(x_i\frac{1-\varepsilon}{n+1} + \varepsilon\lambda^i(x)x_i\right) \\
&= \sum_{i=0}^{n} x_i\left(\frac{1-\varepsilon}{n+1} + \varepsilon\lambda^i(x)\right) = \sum_{i=0}^{n}\alpha^i(x)x_i; \\
\sum_{i=0}^{n}\alpha^i(x) &= 1, \quad \text{where } \alpha^i(x) = \frac{(1-\varepsilon)}{n+1} + \varepsilon\lambda^i(x).
\end{aligned}$$

By the choice of ε_0, we have $\alpha^i(x) \geq 0$ $\forall i$ for all $x \in O_1$ and all $\varepsilon \in (0, \varepsilon_0)$. Therefore $\varepsilon x \in S$ for all above mentioned x and ε. So, $S \supset O_{\varepsilon_0}$ and hence the interior $\operatorname{int} S$ is nonempty. □

Exercise 3.15. *Prove that in the space $\mathbb{R}^n$ vectors $x_0, \dots, x_n$ are affinely independent if and only if* $\operatorname{int}(\operatorname{conv}\{x_0, \dots, x_n\}) \neq \varnothing$.

The relative interior of convex sets

Let X be a normed space and $A \subseteq X$.

Definition 3.16. The relative interior of a convex set A is the interior of A with respect to aff A. Namely, a point x_0 belongs to the relative interior of the set A if there exists such an $\varepsilon > 0$ that

$$O(x_0, \varepsilon) \bigcap \operatorname{aff} A \subseteq A.$$

The relative interior of a set A will be denoted by ri A.

Theorem 3.17. *Let $A \subset \mathbb{R}^n$ be a convex set. Then its relative interior* ri A *is nonempty.*

Proof. Suppose, first, that aff $A = \mathbb{R}^n$. Consider a maximal affinely independent system of points $x_0, \dots, x_m$ from A. It is obvious that $m \le n$. The set A does not lie in any proper linear manifold. Therefore, if $m < n$, then the set A contains a vector which does not belong to aff $(x_0, \dots, x_m)$ and, hence, these points do not form a maximal affine independent system in A. The obtained contradiction shows that $m = n$, implying, by Theorem 3.14 that int $A \ne \varnothing$.

Now, let aff $A \ne \mathbb{R}^n$. Without loss of generality, we may assume that $0 \in$ aff A, i.e., aff A is the linear subspace of a certain dimension r, which may be identified with the space $\mathbb{R}^r$, reducing this case to the one considered above. □

The following example demonstrates that, if the space X is infinite-dimensional, then it may contain a convex set A, whose relative interior ri A is empty.

Example 3.18. Let $X = l_2$ and A is a nonnegative orthant in l_2, i.e., $A = \{x : x^i \ge 0 \ \forall i\}$. It is easy to see that aff $A = l_2$ and, hence, ri A = int A. Let us show that int $A = \emptyset$. In fact, let $x \in A$. Take an arbitrary $\varepsilon > 0$. By virtue of convergence of the series $\sum_{i=1}^{\infty}(x^i)^2$, there exists such a number i, that $0 \le x^i < \varepsilon/2$. Denote by x_ε the sequence obtained from x by the substitution of the element x^i with $(-\varepsilon/2)$. The inequality $|x^i - (-\varepsilon/2)| < \varepsilon$ is evident. So, $|x - x_\varepsilon| < \varepsilon$. Moreover, it is clear that $x_\varepsilon \notin A$. Therefore, $x \notin$ int A and, hence int $A = \emptyset$.

For the relative interior in $\mathbb{R}^n$, the following remarkable formula holds true.

Lemma 3.19. *Let A, B be nonempty convex subsets of $\mathbb{R}^n$. Then*

$$\operatorname{ri}(A + B) = \operatorname{ri} A + \operatorname{ri} B.$$

The proof of this assertion (somewhat nontrivial) may be found in [2], Corollary 6.6.2. Here we notice only that in infinite-dimensional spaces the statement of this lemma may be violated. In fact, let A be the nonnegative orthant in a Hilbert space l_2 and $B = -A$. Then, as demonstrated in Example 3.18, ri A = ri $B = \emptyset$, however $A + B = l_2$ and hence ri $(A + B) = l_2$.

Lemma 3.20. *Let the set A be convex, $x_1 \in$ cl A, $x_2 \in$ ri A. Then*

$$(1 - \alpha)x_1 + \alpha x_2 \in \operatorname{ri} A, \quad \forall \alpha \in (0, 1].$$

Proof. Since $x_2 \in \operatorname{ri} A$, then $(x_2 + \operatorname{lin}(A - x_2) \cap O_\varepsilon) \subseteq A$ for a certain $\varepsilon > 0$. From here and from the convexity of A it follows that

$$\begin{aligned}&(1-\alpha)x_1 + \alpha\,(x_2 + \operatorname{lin}(A - x_2) \cap O_\varepsilon)\\ &\quad = (1-\alpha)x_1 + \alpha x_2 + \operatorname{lin}(A - x_2) \cap (\alpha O_\varepsilon) \subseteq \operatorname{cl} A.\end{aligned}$$

In other words, $(1-\alpha)x_1 + \alpha x_2 \in \operatorname{ri} A$, since $\operatorname{ri} A = \operatorname{ri}(\operatorname{cl} A)$ (the latter can be proved in the same way as Corollary 2.8). □

From Theorem 3.17 and Lemma 3.20 it follows that, if $x_0 \in A \subset \mathbb{R}^n$ and a set A is convex, then there exists a sequence $\{x_i\}$, which lies in $\operatorname{ri} A$ and converges to x_0.

4 Separation theorems for convex sets

Notions of separability of convex sets

Let X be a normed space, A and B its nonempty subsets.

Definition 4.1. Sets A and B may be separated if there exists a continuous linear functional $\ell \neq 0$ such that

$$\sup_{x \in A} \langle \ell, x \rangle \leq \inf_{y \in B} \langle \ell, y \rangle. \tag{4.1}$$

Definition 4.2. Sets A and B may be strictly separated if there exists a continuous linear functional $\ell \neq 0$ such that

$$\sup_{x \in A} \langle \ell, x \rangle < \inf_{y \in B} \langle \ell, y \rangle$$

(i.e., inequality (4.1) is strict).

Here and below the notation $\langle \ell, x \rangle$ means the action of a linear functional ℓ on a vector x.

We will say also that a functional ℓ separates (strictly separates) the sets A and B.

Notice that the separability of sets A and B is equivalent to the existence of a continuous linear functional $\ell \neq 0$ and a number $\gamma \in \mathbb{R}$ such that

$$\langle \ell, x \rangle \leq \gamma \leq \langle \ell, y \rangle \quad \forall x \in A, \ \forall y \in B.$$

Theorems on separability of convex sets play the key role in obtaining new results in convex analysis. For example, in finite-dimensional spaces, they guarantee that each two convex sets with disjoint relative interiors may be separated. If the space is infinite-dimensional, it is necessary to assume, additionally, that the algebraic interior (kernel) of at least one of these sets should be nonempty (the notion of the algebraic interior will be introduce below).

Before we pass to properly separability theorems, let us consider the following useful assertion.

Proposition 4.3. *Sets A and B may be separated if and only if the set $(A - B)$ may be separated from zero.*

Proof. Necessity. Let sets A and B be separated. Then there exists ℓ such that

$$\langle \ell, x \rangle \leq \langle \ell, y \rangle \Rightarrow \langle \ell, x - y \rangle \leq 0 \quad \forall x \in A, \ y \in B,$$

implying

$$\sup_{z \in A-B} \langle \ell, z \rangle \leq 0 = \langle \ell, 0 \rangle.$$

DOI 10.1515/9783110460308-005

Sufficiency. Let the set $(A - B)$ be separated from zero. Then there exists ℓ such that

$$\begin{aligned}\langle \ell, x - y\rangle \leq 0 &\Rightarrow \langle \ell, x\rangle \leq \langle \ell, y\rangle \quad \forall x \in A,\ \forall y \in B\\ &\Rightarrow \sup_{x\in A} \langle \ell, x\rangle \leq \inf_{x\in B} \langle \ell, x\rangle.\end{aligned}$$ □

Separability in finite-dimensional spaces

Suppose $X = \mathbb{R}^n$.

Lemma 4.4. *Let a nonempty set $C \subset \mathbb{R}^n$ be convex closed and $0 \notin C$. Then C may be strictly separated from zero.*

Proof. Take any $c_0 \in C$ and consider the set

$$\tilde{C} = \{x \in C : |x| \leq |c_0|\}.$$

This set is nonempty, convex and compact. Consider the minimization problem

$$|x|^2 \to \min, \quad x \in \tilde{C}.$$

Since the set $\tilde{C}$ is compact, by the Weierstrass[1] theorem the minimum in this problem is achieved at a certain point $x_0 \in \tilde{C}$. Obviously, $|x_0| \leq |x|\ \forall x \in C$.

Take an arbitrary point $x \in C$ and set $\Delta = x - x_0$. Then, by virtue of convexity of C, we have

$$\begin{aligned}&x_0 + \alpha\Delta = x_0(1 - \alpha) + x\alpha \in C \quad \forall \alpha \in [0, 1]\\ &\Rightarrow \langle x_0 + \alpha\Delta, x_0 + \alpha\Delta\rangle \geq \langle x_0, x_0\rangle\\ &\Rightarrow 2\alpha\langle \Delta, x_0\rangle + \alpha^2\langle \Delta, \Delta\rangle \geq 0 \quad \forall \alpha \in [0, 1].\end{aligned}$$

Dividing the obtained inequality over $\alpha > 0$ and passing to the limit as $\alpha \to 0+$, we get the inequality $\langle \Delta, x_0\rangle \geq 0$. So, taking into account that $x_0 \neq 0$, we have $\langle x_0, x\rangle \geq |x_0|^2 > 0$ for all $x \in C$. Therefore, the strict separability by $\ell = x_0$ is proved. □

Lemma 4.5. *Let a nonempty set $C \subset \mathbb{R}^n$ be convex and $0 \notin \operatorname{int} C$. Then C may be separated from zero.*

Proof. Consider the convex closed set $\operatorname{cl} C$. By Corollary 2.8 of Theorem 2.7, zero does not belong to the interior of the set $\operatorname{cl} C$. Therefore zero either does not belong to the set $\operatorname{cl} C$ or belongs to its boundary. Hence, there exists such a sequence, $\{x_i\}$, converging to zero, that $x_i \notin \operatorname{cl} C\ \forall i$. So, $0 \notin C_i := \operatorname{cl} C - x_i$. It is evident that each set C_i is convex and closed. Therefore, by virtue of the above lemma each of the sets C_i may be separated from zero and, hence there exist such $\ell_i \in \mathbb{R}^n$, $\ell_i \neq 0$, that $\langle \ell_i, \xi\rangle \geq 0$ for all $\xi \in C_i$.

1 Karl Weierstrass (1815–1897), a German mathematician.

Norming each of the vectors ℓ_i, we can suppose them to be of unite length. So, passing, if necessary, to a subsequence we obtain that the sequence $\{\ell_i\}$ converges to a certain $\ell \neq 0$.

From the definition of the sets C_i and the Cauchy[2]–Bunyakovskii[3]–Schwarz[4] inequality we get

$$\langle \ell_i, x\rangle + |l_i|\,|x_i| \geq 0 \quad \forall x \in \operatorname{cl} C \; \forall i.$$

Passing in the obtained inequalities to the limit as $i \to \infty$ we have that $\langle \ell, x\rangle \geq 0$ for all $x \in C$ and, hence the set C is separated from zero. □

Theorem 4.6 (On finite-dimensional separability). *Let A and B be nonempty convex subsets $\mathbb{R}^n$ such that their relative interiors* $\operatorname{ri} A$ *and* $\operatorname{ri} B$ *do not intersect. Then the sets A and B may be separated.*

Proof. Set $C = \operatorname{ri} A - \operatorname{ri} B$. Then $0 \notin C$. Hence, by virtue of Lemma 4.5 and Proposition 4.3, the sets $\operatorname{ri} A$ and $\operatorname{ri} B$ may be separated, i.e., there exist $\ell \neq 0$ and $y \in \mathbb{R}$ such that

$$\langle \ell, x\rangle \leq y \leq \langle \ell, y\rangle \quad \forall x \in \operatorname{ri} A, \; \forall y \in \operatorname{ri} B.$$

Let us show that the functional ℓ also separates the sets A and B. In fact, let $x \in A$. Then there exists the sequence of points $\{x_i\}$ lying in $\operatorname{ri} A$ and converging to x. Therefore, $\langle \ell, x_i\rangle \leq y$ for all i. Passing to the limit as $i \to \infty$ we get $\langle \ell, x\rangle \leq y$. Repeating this reasoning for points y of the set B we obtain the desired result. □

Separability in infinite-dimensional spaces

Let X be a normed space.

Theorem 4.7 (On separability). *Let M and N be nonempty convex subsets of X,* $\operatorname{int} M \neq \varnothing$ *and* $(\operatorname{int} M) \cap N = \varnothing$. *Then the sets M and N may be separated.*

Before proving this theorem, let us consider some important constructions from functional analysis. We begin with the following deep result which is the basis for the whole separability theory in infinite-dimensional spaces.

Definition 4.8. A function $p\colon X \to \mathbb{R}$ is called homogeneously convex functional provided

$$p(\lambda x) = \lambda p(x) \;\forall \lambda \geq 0, \quad p(x+y) \leq p(x) + p(y) \;\forall x, y \in X.$$

2 Augustin Louis Cauchy (1789–1857), a French mathematician.
3 Viktor Yakovlevich Bunyakovskii (1804–1889), a Russian mathematician.
4 Karl Hermann Amandus Schwarz (1843–1921), a German mathematician.

Theorem 4.9 (Hahn[5]–Banach). *Let p be a homogeneously convex functional on a real linear space X. Let X_0 be a linear subspace in X on which a linear functional $f_0\colon X_0 \to \mathbb{R}$ is given. Suppose that f_0 is subordinated to a functional p, i.e.,*

$$f_0(x) \le p(x), \quad \forall x \in X_0.$$

Then f_0 may be extended to a linear functional f on X, subordinated to p on the whole X, i.e.,

$$f(x) \le p(x) \quad \forall x \in X.$$

The proof of this classical theorem may be found, e.g., in [27].

The norm in X is the simplest example of a homogeneously convex functional. However, to prove the separability theorem, it is necessary to apply the Hahn–Banach theorem to another homogeneously convex functional, the Minkowski functional.

Definition 4.10. Let $A \subseteq X$ be a convex set and $0 \in \operatorname{int} A$. Then the functional

$$P_A(x) = \inf\{\, r > 0\colon\ x/r \in A \,\}$$

is called the Minkowski functional of the set A.

It is obvious that the functional Minkowski is finite for each x, nonnegative and positively homogeneous. Moreover, if $x \in A$, then $P_A(x) \le 1$, and if $x \notin A$, then $P_A(x) \ge 1$.

Proposition 4.11. *The Minkowski functional is homogeneously convex.*

Proof. Fix arbitrary x_1, x_2. By definition, for each $\varepsilon > 0$ there exist r_1, r_2 such that $P_A(x_i) < r_i < P_A(x_i) + \varepsilon$, $i = 1, 2$. Then $P_A(x_i/r_i) < 1$ and $x_i/r_i \in A$.

Set $r = r_1 + r_2$. By the convexity of A we have

$$\frac{x_1 + x_2}{r} = \left(\frac{r_1}{r}\frac{x_1}{r_1} + \frac{r_2}{r}\frac{x_2}{r_2}\right) \in A \Rightarrow P_A\Big(\frac{x_1 + x_2}{r}\Big) \le 1$$
$$\Rightarrow P_A(x_1 + x_2) \le r = r_1 + r_2 \le P_A(x_1) + P_A(x_2) + 2\varepsilon.$$

By arbitrariness of $\varepsilon > 0$ we obtain the semi-additivity of the Minkowski functional:

$$P_A(x_1 + x_2) \le P_A(x_1) + P_A(x_2).$$

Hence, from its positive homogeneity it follows that the functional P_A is homogeneously convex. □

Exercise 4.12. *Let $y > 0$ and $A = \{x\colon \|x\| \le y\}$. Prove that then $P_A(x) = y^{-1}\|x\|$.*

Decision. For an arbitrary x we have

$$r > \|x\|/y \Rightarrow x/r \in A, \qquad r < \|x\|/y \Rightarrow x/r \notin A.$$

Hence $P_A(x) = \inf\{\, r > 0\colon\ x/r \in A \,\} = \|x\|/y$.

---

5 Hans Hahn (1879–1934), an Austrian mathematician.

Proposition 4.13. *Let zero belong to the interior of a set $A \subset X$. Then there exists such a $c > 0$, that $P_A(x) \le c\|x\| \ \forall x \in X$.*

Proof. By virtue of the assumption there exists such an $\varepsilon > 0$, that $O_\varepsilon \subset A$. But if a set B is contained in A, then, evidently, $P_A(x) \le P_B(x) \ \forall x \in X$. At the same time, as demonstrated above, $P_{O_\varepsilon}(x) \equiv \|x\|/\varepsilon$. Therefore, the desired inequality holds for $c = \varepsilon^{-1}$. □

Let us return to the proof of Theorem 4.7.

Proof. We will assume that $0 \in \operatorname{int} M$ (we may achieve it by the parallel transfer of the sets along an arbitrary vector from $\operatorname{int} M$).

Let $y_0 \in N$. Then $-y_0 \in -N \Rightarrow -y_0 \in \operatorname{int} M - N$. Set $D = \operatorname{int} M - N + y_0$. Then $0 \in \operatorname{int} D$, $0 \notin \operatorname{int} M - N$ and hence $y_0 \notin D$. Consider the Minkowski functional p of the set D. Then $p(y_0) \ge 1$, since $y_0 \notin D$.

Consider the one-dimensional subspace $L = \operatorname{lin}\{y_0\}$. Define the linear functional f_0 on L by the formula: $f_0(\alpha y_0) = \alpha p(y_0) \ \forall \alpha \in \mathbb{R}$. Let us show that, on the one-dimensional subspace L the constructed linear functional f_0 is subordinated to the functional p. Indeed, if $\alpha \ge 0$, then $f_0(\alpha y_0) = \alpha p(y_0) = p(\alpha y_0)$. In case $\alpha < 0$, we have $f_0(\alpha y_0) \le 0 \le p(\alpha y_0)$ and hence $f_0(\alpha y_0) \le p(\alpha y_0) \ \forall \alpha \in \mathbb{R}$. So we proved that on L the linear functional f_0 is subordinated to the homogeneously convex Minkowski functional p.

By the Hahn–Banach theorem, the functional f_0 may be extended on X to a linear functional f, subordinated to p. Hence $f(y) \le p(y) \ \forall y \in X$ and so $f(-y) \le p(-y)$, implying, by the linearity of the functional f that $-p(-y) \le f(y) \ \forall y \in X$. The linear functional f is bounded on a certain neighborhood of zero since $-p(-y) \le f(y) \le p(y) \Rightarrow |f(y)| \le \max\{p(y), p(-y)\} \ \forall y \in X$, and, by virtue of Proposition 4.13 the Minkowski functional p is also bounded on a certain neighborhood of zero, since, as demonstrated above, $0 \in \operatorname{int} D$. Hence, the linear functional f is continuous, since each bounded linear functional on a normed space is continuous.

Further, $f(y_0) = p(y_0) \ge 1$. For each $y \in D$ we have: $p(y) \le 1$ and hence $f(y) \le 1 \le f(y_0)$. Therefore, the constructed linear functional f separates the set $D = \operatorname{int} M - N + y_0$ and the point y_0. Therefore f separates the set $(\operatorname{int} M - N)$ and zero and hence separates the sets $\operatorname{int} M$ and N. This implies that the sets M and N may be separated. □

In real linear spaces (in which the norm is not defined) convex disjoint sets also may be separated, but, to this end, we should use linear functionals, saying nothing about their continuity since in arbitrary linear spaces the notion of continuity is lacking.

Let us formulate the corresponding separability theorem. Let X be a real linear space.

Definition 4.14. We will say that subsets A, B of a real linear space X may be separated if there exists a linear functional $\ell \neq 0$ defined on X such that

$$\sup_{x \in A} \langle \ell, x \rangle \leq \inf_{y \in B} \langle \ell, y \rangle.$$

This notion of separability differs from Definition 4.1 by the absence of the condition of continuity for the linear functional ℓ.

The algebraic interior of a set $M \subseteq X$ (also called *the kernel of* M) is the collection of all points x_0 such that for each $x \in X$ there exists a number $\varepsilon(x) > 0$ with $x_0 + tx \in M$ for all $t\colon |t| \leq \varepsilon(x)$. Notice that in the case of a Banach space X the algebraic interior of a convex subset coincides with its interior, however, if a normed space X is not complete it is not so: the algebraic interior of a set may be nonempty whereas its interior is empty.

Theorem 4.15. *Let M, N be nonempty convex subsets of a real linear space X, the kernel of the set M is nonempty and does not intersect the set N. Then the sets M and N may be separated in the sense of Definition* 4.14.

The validity of this assertion follows from the proof of Theorem 4.7.

Let us clarify the fundamental difference of separability in infinite-dimensional normed spaces from separability in finite-dimensional spaces. By Theorem 4.6 on finite-dimensional separability, if $A \subset X$ is an arbitrary nonempty convex set and the space X is finite-dimensional, then each boundary point of the set A may be separated from A. If the space X is infinite-dimensional, then, by virtue of Theorem 4.7 it is possible only under the additional assumption that the interior of A is nonempty. The following example demonstrates the essence of this supposition.

Example 4.16. In the Hilbert space $X = l_2$ consider the set Π, called "the Hilbert cube". It consists of such sequences $x = (x^1, x^2, \dots)$, for which $|x^i| \leq 2^{-i}\ \forall i$.

The Hilbert cube Π contains zero, is convex and compact, does not lie in any finite-dimensional subspace and has empty (even algebraic) interior (see, e.g., [27]). Let us show that the Hilbert cube cannot be separated from zero in the sense of Definition 4.1.

In fact, suppose that a linear functional $\ell \in l_2$ separates the set Π and zero. Then $\langle \ell, x \rangle \geq 0\ \forall x \in \Pi$. Replacing in this inequality x with the vectors $e_i^\pm$, $i = 1, 2, \dots,$ where $e_i^\pm \in \Pi$ has $\pm 2^{-i}$ on the i-th position and zeros on other ones, we obtain $\ell = 0$, that is the contradiction. So, the Hilbert cube cannot be separated from zero. Notice that the algebraic interior of Π is empty.

As mentioned above, zero belongs to the Hilbert cube. At the same time, in the Hilbert space l_2 exist convex, closed, unbounded, disjoint sets (naturally, having empty algebraic interior) which cannot be separated. An example of such sets is given in [13] (see Example 1.9.2).

Strict separability of convex sets

Again, let X be a normed space.

Theorem 4.17. *Let A and B be nonempty disjoint convex subsets of X, and moreover, A be compact and B be closed. Then the sets A and B may be strictly separated.*

Proof. Let us show that there exists such an $\varepsilon > 0$, that $A_\varepsilon \cap B = \emptyset$. Here $A_\varepsilon = A + O_\varepsilon$ is the open ε-neighborhood of the set A.

Suppose the contrary, i.e., that the intersection $A_\varepsilon \cap B$ is nonempty for each $\varepsilon > 0$. Then for each number i there exist such $a_i \in A$, $b_i \in B$ that

$$\|a_i - b_i\| \to 0, \quad i \to \infty. \tag{4.2}$$

By virtue of compactness of the set A, passing, if necessary, to subsequence, we will assume that $a_i \to a \in A$. Therefore, by virtue of (4.2) and the representation $b_i = a_i + (b_i - a_i)$, the sequence $\{b_i\}$ also converges to a, where $a \in B$ by the closeness of B. This contradicts the condition that the sets A and B are disjoint. It means that the desired $\varepsilon > 0$ exists.

Consider the sets A_ε and B. Both these sets are convex, disjoint and the interior of the first one is nonempty. Therefore, by Theorem 4.7 these sets may be separated and hence there exists such a continuous linear functional $\ell \neq 0$, that

$$\sup_{x \in A_\varepsilon} \langle \ell, x\rangle \le \inf_{y \in B} \langle \ell, y\rangle.$$

But it is easy to see that

$$\sup_{x \in A_\varepsilon} \langle \ell, x\rangle \ge \sup_{x \in A} \langle \ell, x\rangle + \varepsilon\|\ell\|,$$

from which it follows that the functional ℓ strictly separates the sets A and B. □

Under the additional assumption that X is a Hilbert (or even more generally, a reflexive Banach) space, the statement of Theorem 4.17 may be strengthened by the corresponding weakening of the supposition concerning the set A. Namely, in this case it is sufficient to assume that A and B are nonempty disjoint convex closed subsets of X, where A is (only) bounded. Then the sets A and B, as before, may be separated. For the reader familiar with properties of weak convergence, we present the proof of this assertion.

Let us turn to the proof of Theorem 4.17. It is obvious that to our end, it is sufficient to prove the existence of $\varepsilon > 0$ used in its proof. Assuming the contrary, construct the above described sequences $\{a_i\} \subset A$ and $\{b_i\} \subset B$ for which (4.2) holds. But it is known (see, e.g., [30]) that each closed convex set is weakly closed and, moreover, if it is bounded, then under assumptions concerning the space X, this set is weakly sequentially compact. So, passing, if necessary, to a subsequence, we will assume that the sequence $\{a_i\}$ weakly converges to a certain $a \in A$. Then, by virtue of (4.2), the sequence $\{b_i\}$ also weakly converges to a certain b. By (4.2), $a = b$, and $b \in B$ since B is weakly closed. This contradicts the fact that the sets A and B are disjoint and concludes the proof of the existence of desired $\varepsilon > 0$. The further reasoning is the same as during the proof of Theorem 4.17.

Notice that the strengthening of Theorem 4.17 to the case of an arbitrary normed space is impossible. Namely, there exist examples of Banach spaces in which there exist two bounded closed convex sets which are disjoint, but they cannot be separated. Details can be found in [13], § 1.9.

5 Convex functions

The notion of a convex function. Criteria of convexity

Let X be a real linear space. By the symbol $\bar{\mathbb{R}}$ we will denote *the extended real axis*, namely, $\bar{\mathbb{R}} = \mathbb{R} \cup \{-\infty, +\infty\}$, where for the symbols "minus infinity" $-\infty$ and "plus infinity" $+\infty$ the operation of addition with elements $\alpha \in \mathbb{R}$ is defined by the evident rule $\alpha + (-\infty) = -\infty$, $\alpha + (+\infty) = +\infty$. The multiplication by $\alpha > 0$ is defined as $\alpha \times (+\infty) = +\infty$, $\alpha \times (-\infty) = -\infty$. We will suppose also that $0 \times (+\infty) = 0 \times (-\infty) = 0$, $-(-\infty) = +\infty$, and that the expression $+\infty + (-\infty)$ has no mathematical meaning.

We will consider functions $f: X \to \bar{\mathbb{R}}$. With each such function f we may connect the sets:

$$\operatorname{epi} f = \{(x, \alpha) \in X \times \mathbb{R}: f(x) \le \alpha\},$$

which is called *the epigraph of the function f* and

$$\operatorname{dom} f = \{x \in X: f(x) < +\infty\}$$

named *the domain of f*.

Definition 5.1. A function f is called proper if $\operatorname{dom} f \neq \emptyset$ and $f(x) > -\infty\ \forall x$. A function which is not proper is called non-proper.

Definition 5.2. A function f is called convex if its epigraph $\operatorname{epi} f$ is a convex set. A function f is called concave if the function $(-f)$ is convex.

It can be directly verified that the proper function f is convex if and only if

$$f(\alpha x_1 + (1-\alpha)x_2) \le \alpha f(x_1) + (1-\alpha) f(x_2) \quad \forall \alpha \in [0, 1], \tag{5.1}$$

for each x_1, x_2. From here, applying by induction the reasoning similar to that used while proving Proposition 1.5 we obtain (5.1) and, hence, the convexity of a proper function f is equivalent to the fact that for each natural n we have:

$$f\left(\sum_{i=1}^{n} \alpha_i x_i\right) \le \sum_{i=1}^{n} \alpha_i f(x_i), \quad \forall (\alpha_1, \dots, \alpha_n): \sum_{i=1}^{n} \alpha_i = 1, \ \alpha_i \ge 0,$$

for every points $x_1, \dots, x_n$.

This relation is called *the Jensen*[1] *inequality*. Notice that if a function is convex (not necessarily proper), then the Jensen inequality holds for every points $x_1, \dots, x_n$, for which the collection $f(x_1), \dots, f(x_n)$ does not contain the infinities of various signs.

From the Jensen inequality it follows that the domain of a convex function is convex.

1 Johan Ludvig William Valdemar Jensen (1859–1925), a Danish mathematician.

DOI 10.1515/9783110460308-006

We consider functions which may take infinite values for the sake of convenience. For example, let a function f, initially taking only finite values, be defined on a convex set $A \subset X$ and convex on it, i.e.,

$$f(\alpha x_1 + (1-\alpha)x_2) \le \alpha f(x_1) + (1-\alpha)f(x_2) \quad \forall \alpha \in [0,1],\ \forall x_1, x_2 \in A.$$

If we extend f to the whole space X, by setting $f(x) = +\infty$ for all $x \notin A$, we obtain the convex function on X. Therefore we may, without loss of generality, consider only functions defined on the whole space X.

Let us clear up the structure of non-proper functions. In the following, we suppose that X is a normed space.

Proposition 5.3. *Let a convex function f be non-proper. Then*

$$f(x) = -\infty \quad \forall x \in \operatorname{ri}(\operatorname{dom} f).$$

In other words, a non-proper convex function is infinite in all points except, maybe, the points of the relative boundary of its domain.

Proof. Let $x \in \operatorname{ri}(\operatorname{dom} f)$. Since the function f is non-proper, there exists $u \in X$ such that $f(u) = -\infty$. Replacing the normed space X with a linear hull of two vectors u and x, we may assume, without loss of generality, that the space X is finite-dimensional. Since $x \in \operatorname{ri}(\operatorname{dom} f)$, it is easy to verify (details can be found in [2]) that there exist such $\alpha \in (0,1)$ and $y \in \operatorname{dom} f$, that $x = (1-\alpha)u + \alpha y$. By virtue of convexity of the function f we have

$$f(x) \le (1-\alpha)f(u) + \alpha f(y) \Rightarrow f(x) = -\infty.$$ □

Lemma 5.4. *Let f be a convex proper function and $X = \mathbb{R}^n$. Then there exist such $a \in \mathbb{R}^n$, $b \in \mathbb{R}$ that*

$$f(x) \ge \langle a, x\rangle + b \quad \forall x \in X. \tag{5.2}$$

Proof. By Theorem 3.17, there exists a point $x_0 \in \operatorname{ri}(\operatorname{dom} f)$. The value $f(x_0)$ is finite. Hence $y = (x_0, f(x_0) - 1) \notin \operatorname{epi} f$. Therefore, by virtue of the finite-dimensional separation theorem, the point y may be separated from the convex set $\operatorname{epi} f$. Then there exist $a \in \mathbb{R}^n$ and $\beta \in \mathbb{R}$ such that $(a, \beta) \ne 0$, $a \in \operatorname{aff}(\operatorname{dom} f)$,

$$\alpha\beta + \langle a, x\rangle \le \beta(f(x_0) - 1) + \langle a, x_0\rangle \quad \forall (x, \alpha) \in \operatorname{epi} f. \tag{5.3}$$

If we suppose that $\beta > 0$, then the left-hand side of this inequality can be made arbitrary large for $x = x_0$, $\alpha \to +\infty$. So, $\beta \le 0$. If we suppose that $\beta = 0$, then $a \ne 0$ and, by (5.3):

$$\langle a, x\rangle \le \langle a, x_0\rangle \quad \forall x \in \operatorname{dom} f,$$

contrary to $x_0 \in \operatorname{ri}(\operatorname{dom} f)$ (since $a \in \operatorname{aff}(\operatorname{dom} f)$). Therefore, we proved that $\beta < 0$ and hence, without loss of generality we can assume that $\beta = -1$. For $\alpha = f(x)$, $b = f(x_0) - 1 - \langle a, x_0\rangle$ from (5.3) we obtain (5.2). □

We can define some operations over convex functions, whose results are convex functions again. Let us mention the most useful of these operations. Beforehand, notice the following evident assertion.

Lemma 5.5. *Let A be a convex subset of $X \times \mathbb{R}$. Set*

$$f(x) = \begin{cases} \inf\{\mu\colon (x,\mu) \in A\}, & \text{if } \{\mu\colon (x,\mu) \in A\} \neq \emptyset \\ +\infty, & \text{if } \{\mu\colon (x,\mu) \in A\} = \emptyset. \end{cases}$$

Then the function f is convex.

Moreover, let the set A be closed and satisfy the following condition: if $(x, \alpha) \in A$ and $\beta > \alpha$, then $(x, \beta) \in A$. Then $A = \operatorname{epi} f$.

Let $f_1(x), \dots, f_m(x)$ be convex functions on X. Then the function

$$f(x) = \left(\sum_{i=1}^{m} f_i\right)(x) = \sum_{i=1}^{m} f_i(x),$$

called *the sum of functions* $f_1, \dots, f_m$, is obviously convex.

The infimal convolution of functions $f_1, \dots, f_m$ is the function $f_1 \oplus \cdots \oplus f_m$, defined by the formula

$$(f_1 \oplus \cdots \oplus f_m)(x) = \inf\left\{\sum_{i=1}^{m} f_i(x_i)\colon \sum_{i=1}^{m} x_i = x\right\}.$$

The infimal convolution of convex functions is a convex function since it may be achieved from the set $A = \operatorname{epi} f_1 + \cdots + \operatorname{epi} f_m$ by the construction given in Lemma 5.5.

Let $f_\sigma(x)$ be a family of convex functions depending on parameter $\sigma \in \Sigma$, where Σ is an arbitrary set of parameters. The function f, defined by the formula

$$f(x) = \sup\{f_\sigma(x)\colon \sigma \in \Sigma\},$$

is called *the upper bound of the family of functions* f_σ. The upper bound of a family of convex functions is a convex function. This follows immediately from the definition as well as from the fact that its epigraph $\operatorname{epi} f$ coincides with the intersection of convex sets $\operatorname{epi} f_\sigma$.

Consider some examples of convex functions.

For an arbitrary subset $A \subseteq X$, its *indicator function* δ_A is defined by the relation

$$\delta_A(x) = \begin{cases} 0, & x \in A \\ +\infty, & x \notin A. \end{cases}$$

Evident examples of convex functions are represented by the linear function, the indicator function of a convex set A and the function $f(x) = \|x\|$, where X is a normed space. The convexity of such functions as, for example,

$$f(x) = \begin{cases} +\infty, & x \leq 0 \\ -\ln x, & x > 0, \end{cases} \tag{5.4}$$

$f(x) = \exp(x)$, $f(x) = x^2$, $x \in \mathbb{R}$ or, more generally, $f(x) = |x|^{2k}$, $x \in \mathbb{R}^n$, where k is a positive integer, may be deduced from the following *criterion of convexity of smooth functions.*

Theorem 5.6. *Let X be a Euclidean space and a function f is twice continuously differentiable on X. Then the function f is convex if and only if*

$$\frac{\partial^2 f}{\partial x^2}(x) \geq 0 \quad \forall x \in X. \tag{5.5}$$

(Here the nonnegativity of the quadratic form $Q = \frac{\partial^2 f}{\partial x^2}(x)$ means that $\langle Q\xi, \xi\rangle \geq 0$ $\forall \xi \in X$.)

Proof. Necessity. For convenience, introduce the notation $\frac{\partial^2 f}{\partial x^2}(x) = f''(x)$. Suppose that a function f is convex. Fix arbitrary $x, y \in X$. From the definition of convexity, for each $\lambda \in [0, 1]$ we have

$$f(\lambda x + (1-\lambda)y) \leq \lambda f(x) + (1-\lambda)f(y),$$

implying

$$\lambda f(x) \geq \lambda f(y) + [f(y + \lambda(x-y)) - f(y)].$$

For each fixed $\lambda \in [0, 1]$ define the scalar function $\varphi_\lambda : [0, 1] \to \mathbb{R}$ by the formula $\varphi_\lambda(\theta) = f(y + \theta\lambda(x-y))$. By the Lagrange[2] mean value theorem, for each $\lambda \in [0, 1]$ there exists $\theta_\lambda \in [0, 1]$ such that $\varphi_\lambda(1) - \varphi_\lambda(0) = \varphi'_\lambda(\theta_\lambda)$. So, evaluating the derivative φ'_λ, as the derivative of a composite function, we obtain

$$f(y + \lambda(x-y)) - f(y) = \langle f'(y + \theta_\lambda\lambda(x-y)), \lambda(x-y)\rangle.$$

By substituting this expression into the above inequality, we obtain

$$\forall \lambda \in [0, 1]\ \exists \theta_\lambda \in [0, 1] : \lambda f(x) \geq \lambda f(y) + \langle f'(y + \theta_\lambda\lambda(x-y)), \lambda(x-y)\rangle$$

Dividing both parts of this inequality over $\lambda > 0$, for $\lambda \to 0+$ we have

$$f(x) \geq f(y) + \langle f'(y), x - y\rangle.$$

Changing x and y, we obtain

$$f(y) \geq f(x) + \langle f'(x), y - x\rangle.$$

Adding the obtained inequalities we have

$$\langle f'(y) - f'(x), y - x\rangle \geq 0 \quad \forall x, y.$$

2 Joseph Louis Lagrange (1736–1813), a French mathematician.

Let $y = x + \varepsilon h$ and $\varepsilon > 0$. Repeating the above reasonings, based on the Lagrange formula, we obtain that for each $\varepsilon > 0$ there exists $\tilde{\theta}_\varepsilon \in [0, 1]$ such that

$$\langle f''(x + \tilde{\theta}_\varepsilon \varepsilon h)\varepsilon h, \varepsilon h\rangle \geq 0.$$

Dividing both parts of this inequality over $\varepsilon^2 > 0$, for $\varepsilon \to 0+$ we finally get $\langle f''(x)h, h\rangle \geq 0 \ \forall h \in X$, proving (5.5).

Sufficiency. Now, let (5.5) hold. Fix arbitrary $x, y \in X$. Consider the scalar function $\varphi(\alpha) = \langle f'(x + \alpha(y - x)), y - x\rangle$, $\alpha \in [0, 1]$. Applying to this function, as above, the Lagrange formula, we obtain the existence of such $\theta \in [0, 1]$, that

$$\langle f'(y) - f'(x), y - x\rangle = \langle f''(x + \theta(y - x))(y - x), y - x\rangle \geq 0. \tag{5.6}$$

Let $\lambda \in [0, 1]$. Set $z = \lambda x + (1 - \lambda)y$. Then $x - z = (1 - \lambda)(x - y)$, $y - z = \lambda(y - x)$, and from the Newton[3]–Leibniz[4] formula we get

$$\begin{aligned}
&\lambda f(x) + (1 - \lambda)f(y) - f(\lambda x + (1 - \lambda)y) \\
&\qquad = \lambda(f(x) - f(z)) + (1 - \lambda)(f(y) - f(z)) \\
&\qquad = \lambda \int_0^1 \langle f'(z + t(x - z)), x - z\rangle\, dt + (1 - \lambda)\int_0^1 \langle f'(z + t(y - z)), y - z\rangle\, dt \\
&\qquad = \lambda(1 - \lambda)\int_0^1 \langle f'(z + t(x - z)) - f'(z + t(y - z)), x - y\rangle\, dt.
\end{aligned}$$

Set $u = z + t(x - z)$, $v = z + t(y - z)$. By virtue of (5.6)

$$\langle f'(u) - f'(v), u - v\rangle \geq 0.$$

From here we obtain

$$\langle f'(z + t(x - z)) - f'(z + t(y - z)), x - y\rangle \geq 0 \quad \forall t \in [0, 1]$$

and, hence, $\lambda f(x) + (1 - \lambda)f(y) - f(\lambda x + (1 - \lambda)y) \geq 0$, completing the proof of convexity of the function f. □

Corollary 5.7. *If a function f is convex and twice continuously differentiable in a certain neighborhood of a point $x_0 \in X$, then*

$$\frac{\partial^2 f}{\partial x^2}(x_0) \geq 0.$$

---

3 Sir Isaac Newton (1642–1727), an English physicist and mathematician.

4 Gottfried Wilhelm Leibniz (1646–1716), a German mathematician and philosopher.

Example 5.8. Consider the function $f(x) = -|x|$, $x \in \mathbb{R}$. It is continuous, twice continuously differentiable at all points $x \neq 0$ and $f''(x) = 0\ \forall x \neq 0$. So, for the function f, condition (5.5) holds for all x, excepting only one point, and this function is non-convex.

Example 5.9. By virtue of the above criterion, the function (5.4) is convex. So, by the Jensen inequality, for each fixed positive integer n we have:

$$-\ln\left(\sum_{i=1}^{n} \alpha_i x_i\right) \leq -\sum_{i=1}^{n} \alpha_i \ln x_i = -\ln\left(\prod_{i=1}^{n} x_i^{\alpha_i}\right)$$

$$\forall x_1, \dots, x_n > 0,\ \forall \alpha_1, \dots, \alpha_n \geq 0 \colon \sum_{i=1}^{n} \alpha_i = 1.$$

Hence, the following inequality holds

$$\sum_{i=1}^{n} \alpha_i x_i \geq \prod_{i=1}^{n} x_i^{\alpha_i}.$$

For $\alpha_i = 1/n$ it turns into the classical the arithmetic mean and the geometric mean inequality:

$$\frac{x_1 + \dots + x_n}{n} \geq \left(\prod_{i=1}^{n} x_i\right)^{1/n}.$$

Example 5.10. For $p > 1$ consider the function on $\mathbb{R}$:

$$f(x) = \begin{cases} +\infty, & x < 0 \\ x^p, & x \geq 0. \end{cases}$$

By the above convexity criterion, it is convex. Set $q = \frac{p}{p-1}$. Then

$$\frac{1}{p} + \frac{1}{q} = 1.$$

Numbers p and q are called conjugate.

Fix arbitrary positive integer n. For each positive numbers $x_1, \dots, x_n, d_1, \dots, d_n$, the Jensen inequality in points $x_1, \dots, x_n$ for $\alpha_i = \frac{d_i}{d_1 + \dots + d_n}$ has the form

$$\left(\sum_{i=1}^{n} d_i\right)^{-p} \left(\sum_{i=1}^{n} d_i x_i\right)^{p} \leq \sum_{i=1}^{n} d_i x_i^p \left(\sum_{i=1}^{n} d_i\right)^{-1}$$

$$\Rightarrow \sum_{i=1}^{n} d_i x_i \leq \left(\sum_{i=1}^{n} d_i x_i^p\right)^{1/p} \left(\sum_{i=1}^{n} d_i\right)^{1/q}.$$

Take arbitrary $a_i, b_i > 0$. For them, evidently, we may find such $x_i > 0$, $d_i > 0$, that $d_i x_i = a_i b_i$, $d_i x_i^p = a_i^p$, $i = 1, \dots, n$. But then $d_i = b_i^q$ and the above inequality immediately implies the Hölder[5] inequality:

$$\sum_{i=1}^{n} a_i b_i \le \left(\sum_{i=1}^{n} a_i^p\right)^{1/p} \left(\sum_{i=1}^{n} b_i^q\right)^{1/q},$$

which holds for all nonnegative a_i, b_i.

Lemma 5.11. *Let X be a normed space, function $f\colon X \to \mathbb{R}$ continuous and*

$$f\left(\frac{x+y}{2}\right) \le \frac{1}{2}(f(x)+f(y)) \quad \forall x, y \in X,$$

i.e., the Jensen inequality holds only for $\alpha = 1/2$. Then the function f is convex.

Proof. We have

$$\begin{aligned} f\left(\frac{3}{4}x + \frac{1}{4}y\right) &\le \frac{1}{2}\left(f(x) + f\left(\frac{x+y}{2}\right)\right) \\ &\le \frac{1}{2}f(x) + \frac{1}{4}(f(x)+f(y)) = \frac{3}{4}f(x) + \frac{1}{4}f(y). \end{aligned}$$

Similarly,

$$f\left(\frac{1}{4}x + \frac{3}{4}y\right) \le \frac{1}{4}f(x) + \frac{3}{4}f(y).$$

By induction we can prove that for each $\alpha = \frac{m}{2^k} \in (0, 1)$ the Jensen inequality holds $f(\alpha x + (1-\alpha)y) \le \alpha f(x) + (1-\alpha)f(y)$. In fact, if this inequality is true for $\alpha_1, \alpha_2 \in (0, 1)$, then

$$\begin{aligned} f\Big(\frac{1}{2}(\alpha_1 x + (1-\alpha_1)y &+ \alpha_2 x + (1-\alpha_2)y)\Big) \\ &\le \frac{1}{2}(f(\alpha_1 x + (1-\alpha_1)y) + f(\alpha_2 x + (1-\alpha_2)y)) \\ &\le \frac{\alpha_1+\alpha_2}{2}f(x) + \frac{(1-\alpha_1)+(1-\alpha_2)}{2}f(y). \end{aligned}$$

Now, prove that the function f is convex, i.e., the Jensen inequality holds for arbitrary $\alpha, \beta \in [0, 1]$, $\alpha + \beta = 1$. Choose $\{\alpha_n, \beta_n\}$ of the form $\alpha_n = \frac{m}{2^k}$ so, that $\{\alpha_n, \beta_n\} \to \{\alpha, \beta\}$ as $n \to \infty$. From the continuity of f it follows that $f(\alpha_n x + \beta_n y) \to f(\alpha x + \beta y)$ as $n \to \infty$.

From the above it follows

$$f(\alpha_n x + \beta_n y) \le \alpha_n f(x) + \beta_n f(y) \quad \forall n.$$

Tending $n \to \infty$ we get

$$f(\alpha x + \beta y) \le \alpha f(x) + \beta f(y). \qquad \square$$

5 Otto Ludwig Hölder (1859–1937), a German mathematician.

It is known that there exists a scalar function $f: \mathbb{R} \to \mathbb{R}$, for which we have

$$f(x+y) = f(x) + f(y) \quad \forall x, y,$$

however the function f is discontinuous at each point. The construction of this function is based on the Zermelo[6] theorem (see [32]). It is obvious that for this function for each x the following equality holds: $f(2x) = 2f(x)$, from which it follows that

$$f\left(\frac{x+y}{2}\right) = \frac{f(x)+f(y)}{2} \quad \forall x, y.$$

Therefore, this discontinuous function satisfies the Jensen inequality for $\alpha = 1/2$. But this function is not convex, since, as was proved in the previous section, each convex function $f: \mathbb{R}^n \to \mathbb{R}$, taking only finite values, is continuous. Hence the assumption about continuity is essential in the proved lemma.

6 Ernst Friedrich Ferdinand Zermelo (1871–1953), a German mathematician.

6 Closedness, boundedness, continuity, and Lipschitz property of convex functions

As previously, let X be a normed space. As usual, we say that a function $f: X \to \bar{\mathbb{R}}$ is continuous at a point $x_0 \in X$, if for each converging sequence $x_i \to x_0$ we have $f(x_i) \to f(x_0)$ as $i \to \infty$. From here it follows, in particular, that if a function f is continuous at a point $x_0 \in \operatorname{dom} f$, then $x_0 \in \operatorname{int}(\operatorname{dom} f)$.

Definition 6.1. A function $f: X \to \bar{\mathbb{R}}$ is called lower semicontinuous at a point x_0, if

$$\varliminf_{x_i \to x_0} f(x_i) \geq f(x_0).$$

A function f is called upper semicontinuous at a point x_0, if the function $(-f)$ is lower semicontinuous. A function is lower semicontinuous [upper semicontinuous] if it is lower semicontinuous [resp., upper semicontinuous] at all points of X.

Definition 6.2. A function is called closed if its epigraph is closed.

Let a function f be given. Consider its epigraph $\operatorname{epi} f$ and denote by A its closure. Define the function $\bar{f}$ by the formula $\bar{f}(x) = \inf\{\alpha: (x, \alpha) \in A\}$. Then $\operatorname{epi} \bar{f} = A = \operatorname{cl}(\operatorname{epi} f)$. Naturally, the function $\bar{f}$ is closed. It is called *the closure of function f* and denoted by $\operatorname{cl} f$. It is obvious that

$$\operatorname{epi}(\operatorname{cl} f) = \operatorname{cl}(\operatorname{epi} f), \quad (\operatorname{cl} f)(x) \leq f(x) \quad \forall x \in X. \tag{6.1}$$

Proposition 6.3. *Let f be a convex proper function and $X = \mathbb{R}^n$. Then its closure $\operatorname{cl} f$ is also proper function.*

Proof. The validity of this assertion immediately follows from Lemma 5.4 and inequality (6.1). □

For a function f, the sets of the form

$$\mathcal{L}_a f = \{x \in X: f(x) \leq a\}.$$

are called Lebesgue[1] sets.

From the Jensen inequality it follows that all Lebesgue sets of a convex function are convex. The converse is not true. For example, for each $n \geq 2$ all Lebesgue sets of the function $f_n(x) = |x|^{1/n}$ are intervals and, hence, convex. But, in virtue of the criterion of convexity, given in the previous section, each of functions f_n is not convex.

Lemma 6.4. *A function f is lower semicontinuous if and only if for each real a, its Lebesgue set $\mathcal{L}_a f$ is closed.*

1 Henri Leon Lebesgue (1875–1941), a French mathematician.

DOI 10.1515/9783110460308-007

Proof. Necessity. In fact, let f be lower semicontinuous. Consider a limit point x_0 of a sequence $\{x_k\} \subset \mathcal{L}_a f$. Then, by definition $f(x_k) \le a\ \forall k$. By virtue of lower semicontinuity we have

$$f(x_0) \le \varliminf_{x_k \to x_0} f(x_k) \le a \Rightarrow f(x_0) \le a \Rightarrow x_0 \in \mathcal{L}_a f,$$

justifying the closedness of the Lebesgue set $\mathcal{L}_a f$.

Sufficiency. Suppose that there exists a point x_0, at which the function f is not lower semicontinuous. Then there exists such a sequence $x_k \to x_0$ that

$$b = \lim_{x_k \to x_0} f(x_k) < f(x_0).$$

(The case $b = -\infty$ is not excluded.) Choose a number a so that $b < a < f(x_0)$. Then $x_k \in \mathcal{L}_a f$ for all sufficiently large k, from here, by virtue of the closedness $\mathcal{L}_a f$, we have $x_0 \in \mathcal{L}_a f$. But, by construction, $f(x_0) > a$. The contradiction completes the reasonings. □

Lemma 6.5. *A function is closed if and only if it is lower semicontinuous.*

Proof. Necessity. Let a function f be closed. Consider a Lebesgue set $\mathcal{L}_a f$ and a sequence of its points $\{x_i\}$, converging to a point x_0. Then

$$f(x_i) \le a \Rightarrow (x_i, a) \in \operatorname{epi} f, \quad (x_i, a) \to (x_0, a).$$

From the definition of the closedness of f it follows that $(x_0, a) \in \operatorname{epi} f \Rightarrow f(x_0) \le a \Rightarrow x_0 \in \mathcal{L}_a f$. So, all Lebesgue sets of the function f are closed, implying its lower semicontinuity.

Sufficiency. Let a function f be lower semicontinuous. Consider a sequence of points $(x_i, a_i) \in \operatorname{epi} f$ and let $(x_i, a_i) \to (x_0, a_0)$, $i \to \infty$. To prove the closedness of the function f, we need to show that $(x_0, a_0) \in \operatorname{epi} f$. Suppose the contrary, i.e., that $(x_0, a_0) \notin \operatorname{epi} f$. Then $f(x_0) > a_0$.

Choose a number y such that $a_0 < y < f(x_0)$. Then $x_0 \notin \mathcal{L}_y f$ and, hence, there exists $\varepsilon > 0$ such that $O(x_0, \varepsilon) \cap \mathcal{L}_y f = \varnothing$ since the Lebesgue sets of a lower semicontinuous function are closed. Then, for all sufficiently large i we have $x_i \notin \mathcal{L}_y f$, yielding $f(x_i) > y$. Since $y > a_0$, the inequality $f(x_i) > a_i$ is true for all sufficiently large i, contrary to $(x_i, a_i) \in \operatorname{epi} f$. □

Therefore, we demonstrated that the following three properties of a function are equivalent: the lower semicontinuity, the closedness of all its Lebesgue sets, and the closedness of the function itself. Now, let us clear up when a convex function is upper semicontinuous.

A set M will be called *symplectic* if it may be represented as the union of a finite number of simplexes.

Proposition 6.6. *Let $f : \mathbb{R}^n \to \bar{\mathbb{R}}$ be a convex function and $M \subseteq \operatorname{dom} f$ a symplectic set. Then the restriction of the function f to M is upper semicontinuous (i.e., if a sequence $\{x_i\}$ belongs to the set M and converges to a point x_0, then the upper limit of the sequence $\{f(x_i)\}$ does not exceed $f(x_0)$).*

Proof. It is sufficient to consider the case when M is an n-dimensional simplex. Take a point $x_0 \in M$ and prove that the restriction of f is upper semicontinuous at this point. Make a barycentric decomposition (triangulation) of the simplex M in the following way. Take any $(n-1)$-dimensional face Γ_i of the simplex M (i.e., the convex hull of n its vertexes) and set $M_i = \operatorname{conv}(\Gamma_i \cup \{x_0\})$. Therefore, the initial simplex will be represented as the union of $(n+1)$ simplexes M_i, each of them has x_0 as the vertex and it is sufficient to prove that f is upper semicontinuous on each of these simplexes. So, we can assume, without loss of generality, that the initial simplex M has the form $M = \operatorname{conv}\{x_0, \dots, x_n\}$. Also, for convenience, let us take $x_0 = 0$.

An arbitrary point $x \in M$ can be uniquely represented in the form $x = \sum_{i=0}^{n} \alpha^i(x)x_i$, where $\alpha(x) = (\alpha^0(x), \dots, \alpha^n(x))$ are barycentric coordinates of the point x. By Theorem 3.12 the map α is continuous. Hence, for $x \to 0$ we have $\alpha^0(x) \to 1$, $\alpha^i(x) \to 0$, $i = 1, \dots, n$. By virtue of convexity of the function f we get

$$
\begin{aligned}
f(x) = f\left(\sum_{i=0}^{n} \alpha^i(x)x_i\right) &\le \sum_{i=0}^{n} \alpha^i(x)f(x_i) \\
&= \alpha^0(x)f(0) + \sum_{i=1}^{n} \alpha^i(x)f(x_i),
\end{aligned}
$$

where, taking into account the adopted agreement, for those numbers i, for which $\alpha^i(x) = 0$, $f(x_i) = -\infty$, we set $\alpha^i(x)f(x_i) = 0$. From the obtained inequality we conclude that the upper limit of the function f as $x \to 0$, $x \in M$ is no greater than $f(0)$, concluding the proof. □

An interval on the line $\mathbb{R}$ is the symplectic set. Hence, if a convex function $f\colon \mathbb{R} \to \bar{\mathbb{R}}$ takes only finite values on a certain interval, then it is upper semicontinuous on it. Nevertheless, if $f\colon \mathbb{R}^n \to \bar{\mathbb{R}}$ is a convex function, but a set $M \subset \operatorname{dom} f$ is not symplectic, the restriction of f to M may be not upper semicontinuous. Here is the corresponding example.

Example 6.7. Let $M = \{(x^1, x^2)\colon (x^1)^2 + (x^2)^2 \le 1\}$ be a unit disc in $\mathbb{R}^2$. Then M is a convex set which is not symplectic. Introduce a parametrization on the boundary of the disc $\partial M = \{(x^1, x^2) = (\cos\varphi, \sin\varphi), \varphi \in [0, 2\pi)\}$. Define a convex function f:

$$
f(x) = \begin{cases} 0, & x \in \operatorname{int} M \\ \varphi, & x = (\cos\varphi, \sin\varphi) \in \partial M \\ +\infty, & x \notin M. \end{cases}
$$

Then for the sequence of points

$$
x_n = \left(\cos\left(2\pi - \frac{1}{n}\right), \sin\left(2\pi - \frac{1}{n}\right)\right) \in \partial M, \quad x_n \to x_0 = (1, 0) \text{ as } n \to \infty,
$$

we get:

$$
\lim_{n\to\infty} f(x_n) = \lim_{n\to\infty}\left(2\pi - \frac{1}{n}\right) = 2\pi > f(x_0) = 0.
$$

This shows that the restriction of the considered function to M is not upper semicontinuous at the point $x_0 = (1, 0)$.

Theorem 6.8. *Let a convex proper function $f : X \to \bar{\mathbb{R}}$ be bounded above in a certain neighborhood of a given point x_0. Then f is continuous in this neighborhood.*

Proof. Without loss of generality, suppose that $x_0 = 0$ and $f(0) = 0$. Choose numbers $c > 0$ and $\delta > 0$ so that $f(x) \le c\ \forall x : \|x\| \le \delta$.

It is sufficient to prove that for an arbitrary $x : 0 < \|x\| \le \delta$ we have

$$|f(x)| \le \frac{c}{\delta}\|x\|. \tag{6.2}$$

Let us do it.

Fix an arbitrary $x \neq 0$, $\|x\| \le \delta$ and set $\varepsilon = \frac{c}{\delta}\|x\|$. So, $0 < \frac{\varepsilon}{c} \le 1$ and $\|\frac{c}{\varepsilon}x\| = \delta$. By virtue of the convexity of f, representing

$$x = \frac{\varepsilon}{c}\left(\frac{c}{\varepsilon}x\right) + \left(1 - \frac{\varepsilon}{c}\right)0,$$

we get

$$\begin{aligned} f(x) &\le \frac{\varepsilon}{c} f\left(\frac{c}{\varepsilon}x\right) + \left(1 - \frac{\varepsilon}{c}\right) f(0) \\ \Rightarrow f(x) &\le \frac{\varepsilon}{c} f\left(\frac{c}{\varepsilon}x\right) \le \frac{\varepsilon}{c} c = \varepsilon \Rightarrow f(x) \le \varepsilon. \end{aligned}$$

Further,

$$0 = \frac{1}{1 + \varepsilon/c} x + \frac{\varepsilon/c}{1 + \varepsilon/c}\left(-\frac{c}{\varepsilon}x\right).$$

From here, by the convexity of the function f we have

$$\begin{aligned} 0 = f(0) &\le \frac{1}{1 + \varepsilon/c} f(x) + \frac{\varepsilon/c}{1 + \varepsilon/c} f\left(-\frac{c}{\varepsilon}x\right) \\ &\le \frac{1}{1 + \varepsilon/c} f(x) + \frac{\varepsilon}{1 + \varepsilon/c} \Rightarrow f(x) \ge -\varepsilon. \end{aligned}$$ □

The proved theorem has a few important corollaries. Consider some of them. In fact, we proved a more strong assertion than the continuity of the function f in a neighborhood of the considered point. Namely, the next statement immediately follows from the proof of the theorem.

Corollary 6.9. *Let for a convex proper function $f : X \to \bar{\mathbb{R}}$ and a point $x_0 \in X$ there exist such $c > 0$, $\delta > 0$, that $f(x) \le f(x_0) + c\ \forall x \in O(x_0, 3\delta)$. Then the function f satisfies on the set $O(x_0, \delta)$ the Lipschitz[2] condition with a constant c/δ, i.e.,*

$$|f(x_2) - f(x_1)| \le \frac{c}{\delta}\|x_2 - x_1\| \quad \forall x_1, x_2 \in O(x_0, \delta).$$

2 Rudolf Otto Sigismund Lipschitz (1832–1903), a German mathematician.

Proof. Set $\tilde{f}(x) = f(x) - f(x_0)$. From inequality (6.2) we obtain that also $-\tilde{f}(x') \le c \; \forall x' : \|x' - x_0\| \le \delta$. Take arbitrary $x_1, x_2 \in O(x_0, \delta)$. Then $\tilde{f}(x) \le \tilde{f}(x_1) + 2c \; \forall x \in O(x_1, 2\delta) \subset O(x_0, 3\delta)$ and $x_2 \in O(x_1, 2\delta)$. Applying the previous theorem we get the desired inequality. □

Corollary 6.10. *Let a convex proper function $f : X \to \bar{\mathbb{R}}$ be bounded above on a certain nonempty open set. Then it is continuous on the set* $\operatorname{int}(\operatorname{dom} f) \ne \emptyset$.

Proof. Obviously, there exists such a point x_0, that the function f is bounded above on a certain neighborhood of the point x_0. Take such $\delta > 0$, $c > 0$, that $f(x) \le c \; \forall x \in O(x_0, \delta)$.

Take an arbitrary point $u \in \operatorname{int}(\operatorname{dom} f)$. It is easy to show that there exist such $\gamma \in (0, 1)$, $v \in \operatorname{int}(\operatorname{dom} f)$, that $\gamma x_0 + (1 - \gamma)v = u$. Let us prove that the function f is bounded above in $(\gamma\delta)$-neighborhood of the point u. Indeed, let $x \in O(u, \gamma\delta)$. Then there exists such a $\xi \in X$, that $\|\xi\| \le \delta$ and $x = u + \gamma\xi = \gamma(x_0 + \xi) + (1 - \gamma)v$. From here, by virtue of convexity of the function f we have

$$f(x) \le \gamma f(x_0 + \xi) + (1 - \gamma)f(v) \le c\gamma + (1 - \gamma)f(v),$$

that yields the boundedness of f in $(\gamma\delta)$-neighborhood of the point u. The continuity of the function f at a point u follows from Theorem 6.8. □

If the normed space X is infinite-dimensional, then the assumptions on the boundedness above of a convex function on any nonempty open set are essential. Namely, in the following example we present a class of convex functions defined on an infinite-dimensional space, taking only finite values and nonbounded above on each nonempty open set. These functions are discontinuous at each point.

Example 6.11. It is known that a linear nonbounded functional l exists on each infinite-dimensional normed space. The construction of such a functional is based on the existence of the Hamel[3] base in each linear space and we are not giving it here. Obviously, function $f = l$ is the desired one, i.e., it is convex, takes only finite values, but unbounded (neither above, nor below) on each nonempty open set and discontinuous at each point.

But all this is not so in a finite-dimensional space. The point is that if $f : \mathbb{R}^n \to \bar{\mathbb{R}}$ is a convex function and $x_0 \in \operatorname{int}(\operatorname{dom} f)$, then f is bounded above on a certain neighborhood of the point x_0. Indeed, choose in $\mathbb{R}^n$ such an n-dimensional simplex

$$S = \operatorname{conv}\{a_0, \ldots, a_n\} \subset \operatorname{int}(\operatorname{dom} f),$$

that x_0 is its barycenter, i.e., the barycentric coordinates of the point x_0 have the form $(1/(n + 1), \ldots, 1/(n + 1))$. This, obviously, can be done. Then, for an arbitrary $x \in S$,

3 Georg Karl Wilhelm Hamel (1877–1954), a German mathematician.

by virtue of the Jensen inequality we have

$$f(x) = f\left(\sum_{i=0}^{n} \alpha_i a_i\right) \le \sum_{i=0}^{n} \alpha_i f(a_i),$$

where $(\alpha_0, \dots, \alpha_n)$ are barycentric coordinates of the point x. From this inequality (since $0 \le \alpha_i \le 1\ \forall i$) it follows that f is bounded above on the simplex S, and hence, on a certain neighborhood of the point $x_0 \in \operatorname{int} S$.

Therefore, by virtue of the above mentioned and Theorem 6.8, a convex proper function $f: \mathbb{R}^n \to \bar{\mathbb{R}}$ is continuous on the set $\operatorname{int}(\operatorname{dom} f)$. In particular, if a convex function is defined on $\mathbb{R}^n$ and takes only finite values, then it is continuous on the whole $\mathbb{R}^n$.

Theorem 6.12. *Let $f: \mathbb{R}^n \to \bar{\mathbb{R}}$ be a proper convex function and S a convex compact set and $S \subset \operatorname{int} \operatorname{dom} f$. Then f satisfies the Lipschitz condition on the set S.*

Proof. The validity of this assertion can be deduced from Corollary 6.9, but we will give the "direct" proof. First, let us show that there exists $\varepsilon > 0$ such that $(S + B) \subset \operatorname{int}(\operatorname{dom} f)$, where $B = \{x : |x| \le \varepsilon\}$. In fact, supposing the contrary, choose such a sequence $\{x_i\}$, that $x_i \in (O_{1/i} + S)$, $x_i \notin \operatorname{int}(\operatorname{dom} f)$. This sequence is obviously bounded. Therefore, passing, if necessary to a subsequence, we can assume that $x_i \to x_0$. It is evident that $x_0 \in S$, $x_0 \notin \operatorname{int}(\operatorname{dom} f)$, contradicting to the assumption $S \subset \operatorname{int} \operatorname{dom} f$. The existence of the desired $\varepsilon > 0$ is proved.

The set $D = B + S$ is compact. By virtue of the above mentioned, the proper convex function f is continuous on the compact $D \subset \operatorname{int}(\operatorname{dom} f)$. Therefore the function f achieves on D its maximal and minimal values. Set

$$m = \min_{x \in D} f(x), \quad M = \max_{x \in D} f(x).$$

Take arbitrary $x, y \in S$. For them, set $z = y + \frac{y-x}{|y-x|}\varepsilon$. Then $z \in D = S + B$ and $y = (1-\lambda)x + \lambda z$, where $\lambda = \frac{|y-x|}{|y-x|+\varepsilon} < 1$. By virtue of convexity of the function f we have

$$\begin{aligned}
f(y) &\le (1-\lambda)f(x) + \lambda f(z) = f(x) + \lambda(f(z) - f(x)) \\
&\Rightarrow f(y) - f(x) \le \lambda(f(z) - f(x)) = \frac{|y-x|(f(z) - f(x))}{|y-x| + \varepsilon} \\
&\le \frac{|y-x|(M-m)}{|y-x| + \varepsilon} \le \frac{M-m}{\varepsilon}|y-x| \Rightarrow f(y) - f(x) \le c|y-x|,
\end{aligned}$$

where $c = \frac{M-m}{\varepsilon}$. Replacing x with y, we get $f(x) - f(y) \le c|y-x|$ and finally

$$|f(y) - f(x)| \le c|y-x| \quad \forall x, y \in S.$$

We proved that the function f satisfies the Lipschitz condition on the set S with the constant $c = \frac{M-m}{\varepsilon}$. □

Let us make a remark. Above, in all assertions about the continuity and the Lipschitz property of convex functions we assumed that they are proper. However, this assumption is not burdensome. The point is that, by virtue of Proposition 5.3, if a convex function is not proper, then $f(x) = -\infty\ \forall x \in \operatorname{ri}(\operatorname{dom} f)$ and, hence, it is constant on an open set $\operatorname{int}(\operatorname{dom} f)$. For a function taking a constant (even equal to $-\infty$) value, the question on continuity or Lipschitzness loses its sense (it is natural to consider a function identically equal to $-\infty$ on a certain set as continuous and Lipschitz on it).

7 Conjugate functions

The notion of a conjugate function

In this section, we will assume that X is a Hilbert space.

Definition 7.1. Let a function $f: X \to \mathbb{R}$ be given. The Young[1]–Fenchel[2] transform of a function f or the function conjugate to f is called the function defined by the formula

$$f^*(x^*) = \sup_{x \in X} (\langle x^*, x\rangle - f(x)).$$

Notice that if X is a linear normed space then the conjugate function f^* is defined on a conjugate space X^* of linear continuous functionals x^* on X. In the case of a Hilbert space, X and X^* are isomorphic as Euclidean spaces (i.e., with the invariance of the scalar product) and therefore we are identifying them. Nevertheless we leave x^* as the notation of the argument for the conjugate function.

Let us give some examples of conjugate functions. For an affine function $f(x) = \langle a, x\rangle + b$, the conjugate function evidently can be evaluated by the formula

$$f^*(x^*) = \begin{cases} -b, & x^* = a \\ +\infty, & x^* \neq a \end{cases}$$

Exercise 7.2. *For an arbitrary convex function f the multiplication by a positive scalar $\lambda > 0$ is defined by the relation*

$$(\lambda f)(x) = \lambda f(x), \quad x \in X.$$

Prove that

$$(\lambda f)^*(x^*) \equiv \lambda f^*(x^*/\lambda), \quad x^* \in X.$$

Exercise 7.3. *Consider the scalar function $f(x) = p^{-1}|x|^p$ on $\mathbb{R}$, where $p > 1$ is given. To construct f^* let us find the maximum in x of the function $xx^* - p^{-1}|x|^p$. Equating its derivative to zero, we obtain that the maximum of this function is achieved at the point $x = \operatorname{sgn}(x^*)|x^*|^{1/(p-1)}$, where, as usual, $\operatorname{sgn}(a)$ denotes the sign of a number a. Prove that*

$$f^*(x^*) = q^{-1}|x^*|^q.$$

Here p and q are conjugate numbers, i.e., $q^{-1} + p^{-1} = 1$.

Exercise 7.4. *Consider the linear–quadratic function $f(x) = \langle Ax, x\rangle + \langle b, x\rangle + c$ on $\mathbb{R}^n$, where A is a quadratic $n \times n$ symmetric positively defined matrix, $b \in \mathbb{R}^n$ is a given vector.*

---

1 William Henry Young (1863–1942), an English mathematician.

2 Werner Fenchel (1905–1988), a Danish mathematician.

DOI 10.1515/9783110460308-008

Calculate the conjugate function f^* and show that

$$f^*(x^*) = \frac{1}{4}\langle x^* - b, A^{-1}(x^* - b)\rangle - c.$$

Exercise 7.5. *Consider the function*

$$f(x) = \begin{cases} -(a^2 - x^2)^{1/2}, & |x| \le a \\ +\infty, & |x| > a \end{cases}$$

on $\mathbb{R}$, where $a > 0$ is given. Prove that the function f is convex, calculate its conjugate and show that

$$f^*(x^*) = a(1 + (x^*)^2)^{1/2}.$$

Notice that, by definition of a conjugate function,

$$f^*(0) = \sup_x(-f(x)) = -\inf_x f(x).$$

Therefore, the knowledge of the value of a conjugate function at zero is equivalent to the calculation of its greatest lower bound.

Properties of conjugate functions

Proposition 7.6. *A conjugate function is convex and closed.*

Proof. By definition, f^* is the upper bound of the family depending on parameter x, affine (and, hence, convex and continuous) functions $\varphi_x(x^*) = \langle x, x^*\rangle - f(x)$. Therefore the epigraph of the conjugate function f^*, being the intersection in $x \in X$ of the convex closed epigraphs epi φ_x, is the convex and closed set. □

Therefore, the operation which assigns to each convex function its conjugate, defines the mapping from the set of all convex functions to its subset, consisting of closed functions.

Proposition 7.7. *For a convex function f, we have $(\mathrm{cl}\, f)^* = f^*$.*

Proof. As mentioned above, $\mathrm{cl}\,(\mathrm{epi}\, f) = \mathrm{epi}\,(\mathrm{cl}\, f)$. Hence for an arbitrary x^* we have

$$\begin{aligned} f^*(x^*) &= \sup\{(\langle x^*, x\rangle - \alpha) : (x, \alpha) \in \mathrm{epi}\, f\} \\ &= \sup\{(\langle x^*, x\rangle - \alpha) : (x, \alpha) \in \mathrm{cl}\,(\mathrm{epi}\, f)\} \\ &= \sup\{(\langle x^*, x\rangle - \alpha) : (x, \alpha) \in \mathrm{epi}\,(\mathrm{cl}\, f)\} = (\mathrm{cl}\, f)^*(x^*). \end{aligned}$$ □

The definition of a conjugate function yields *the Young–Fenchel inequality*

$$f^*(x^*) + f(x) \ge \langle x, x^*\rangle \quad \forall x, x^* \in X.$$

The second conjugate function f^{**} is defined by the formula $f^{**} = (f^*)^*$.

Proposition 7.8. *For each function f, the following inequality holds*

$$f^{**}(x) \le f(x) \quad \forall x \in X.$$

Proof. By virtue of the Young–Fenchel inequality for each x we have

$$\begin{aligned} & f(x) \ge \langle x, x^* \rangle - f^*(x^*) \quad \forall x^* \in X \\ \Rightarrow & f(x) \ge \sup_{x^*} \{ \langle x, x^* \rangle - f^*(x^*) \} = f^{**}(x). \end{aligned}$$

□

Lemma 7.9. *Let f be a convex, closed and proper function. Then f^* is also a proper function.*

Proof. Let us prove that $f^*(x^*) > -\infty$, $\forall x^* \in X$. Take $x_0 \in \operatorname{dom} f \neq \varnothing$. Then $f^*(x^*) \ge \langle x_0, x^* \rangle - f(x_0) > -\infty$, since $f(x_0) < +\infty$. It remains to prove the existence of a vector $y^* \in X$ for which $f^*(y^*) < +\infty$.

It is obvious that the point $(x_0, f(x_0) - 1)$ does not belong to the closed convex set $\operatorname{epi} f$. Hence, by the separability theorem it may be strictly separated from the convex closed set $\operatorname{epi} f$. Hence there exist $y^* \in X$ and $\beta \in \mathbb{R}$, such that

$$\sup_{(x,\alpha) \in \operatorname{epi} f} \{ \beta\alpha + \langle y^*, x \rangle \} < \beta(f(x_0) - 1) + \langle y^*, x_0 \rangle. \tag{7.1}$$

Let us prove that $\beta < 0$. Indeed, suppose the contrary. The case $\beta > 0$ is impossible since $(x_0, \alpha) \in \operatorname{epi} f$ $\forall \alpha \ge f(x_0) \neq +\infty$ and, hence, for $\beta > 0$ we have $\sup_{(x_0,\alpha) \in \operatorname{epi} f} \beta\alpha = +\infty$, in contradiction to inequality (7.1).

Now, let $\beta = 0$. Then $\sup_{(x,\alpha) \in \operatorname{epi} f} \langle y^*, x \rangle < \langle y^*, x_0 \rangle$, although

$$(x_0, f(x_0)) \in \operatorname{epi} f \Rightarrow \sup_{(x,\alpha) \in \operatorname{epi} f} \langle y^*, x \rangle \ge \langle y^*, x_0 \rangle.$$

The obtained contradiction shows that $\beta < 0$. Therefore, by virtue of positive homogeneity of inequality (7.1) with respect to the variable (y^*, β), we will assume, without loss of generality, that $\beta = -1$.

By virtue of (7.1) we have

$$\begin{aligned} f^*(y^*) &= \sup_x \{ -f(x) + \langle y^*, x \rangle \} = \sup_{(\alpha,x) \in \operatorname{epi} f} \{ -\alpha + \langle y^*, x \rangle \} < \\ &< -(f(x_0) - 1) + \langle y^*, x_0 \rangle \Rightarrow f^*(y^*) < +\infty. \end{aligned}$$

Hence, the function f^* is proper. □

Theorem 7.10 (Fenchel–Moreau[3]). *Let a function f be convex, closed and proper. Then $f^{**} = f$.*

3 Jean Jacques Moreau (1923–2014), a French mathematician.

Proof. It has been proved already that $f^{**} \leq f$. It remains to show that $f^{**} \geq f$.

Suppose the contrary. Then there exists $x_0 \in X$, for which $f^{**}(x_0) < f(x_0)$. Therefore the point $(x_0, f^{**}(x_0))$ is strictly separated from the convex closed set $\operatorname{epi} f$. Hence, there exist $y^* \in X$ and $\beta \in \mathbb{R}$, such that

$$\beta f^{**}(x_0) + \langle y^*, x_0 \rangle > \sup_{(y,\alpha) \in \operatorname{epi} f} (\beta\alpha + \langle y^*, y \rangle). \tag{7.2}$$

Let us prove that $\beta < 0$. Indeed, the case $\beta > 0$ is impossible, which can be justified in the same way as during the proof of Lemma 7.9 (taking into account that $\operatorname{dom} f \neq \varnothing$).

Now, let $\beta = 0$. Then

$$\gamma = \langle y^*, x_0 \rangle - \sup_{y \in \operatorname{dom} f} \langle y^*, y \rangle > 0.$$

By virtue of Lemma 7.9, the function f^* is proper. Therefore there exists $y_1^* \in \operatorname{dom} f^* \neq \varnothing$. For $t > 0$ we have

$$\begin{aligned} f^*(y_1^* + ty^*) &= \sup_{y \in \operatorname{dom} f} (\langle y_1^* + ty^*, y \rangle - f(y)) \\ &\leq \sup_{y \in \operatorname{dom} f} (\langle y_1^*, y \rangle - f(y)) + t \sup_{y \in \operatorname{dom} f} \langle y^*, y \rangle \\ &= f^*(y_1^*) + t \sup_{y \in \operatorname{dom} f} \langle y^*, y \rangle. \end{aligned}$$

From here, by virtue of the Young–Fenchel inequality for the function f^*, it follows that

$$\begin{aligned} f^{**}(x_0) &\geq \langle y_1^* + ty^*, x_0 \rangle - f^*(y_1^* + ty^*) \\ &\geq \langle y_1^*, x_0 \rangle + t\langle y^*, x_0 \rangle - f^*(y_1^*) - t \sup_{y \in \operatorname{dom} f} \langle y^*, y \rangle \\ &= \langle y_1^*, x_0 \rangle - f^*(y_1^*) + t\gamma, \quad \forall t > 0. \end{aligned}$$

We obtained the contradiction, since $\gamma > 0$ and, hence for large t the value $t\gamma$ can be made arbitrarily large and, hence, for sufficiently large $t > 0$ the last inequality does not hold.

Therefore, it is proved that $\beta < 0$ and, hence, without loss of generality, we can assume that $\beta = -1$. By virtue of inequality (7.2) we have

$$-f^{**}(x_0) + \langle y^*, x_0 \rangle > \sup_{y \in \operatorname{dom} f} (-f(y) + \langle y^*, y \rangle) = f^*(y^*),$$

from which

$$\langle y^*, x_0 \rangle > f^*(y^*) + f^{**}(x_0),$$

that contradicts to the Young–Fenchel inequality for the function f^*. The obtained contradiction proves that $f^{**} \geq f$ and, hence, $f^{**} = f$. □

Infimal convolution and the operation of conjugation

Above we defined the operation of infimal convolution of several convex functions. Let us clear up the connection of infimal convolution to the operation of conjugation.

Below we will formulate all assumptions for n functions $f_1, \dots, f_n$ defined on a Hilbert space X, however, for simplicity, all proofs will be presented only for the case of two functions, i.e., for $n = 2$. Notice that for two functions f_1 and f_2 the formula of infimal convolution has the form

$$(f_1 \oplus f_2)(x) = \inf_{x_1+x_2=x} \{f_1(x_1) + f_2(x_2)\} = \inf_{x_1} \{f_1(x_1) + f_2(x - x_1)\}.$$

Lemma 7.11. *For any functions $f_1, \dots, f_n$ we have*

$$(f_1 \oplus f_2 \oplus \cdots \oplus f_n)^* = f_1^* + f_2^* + \cdots + f_n^*.$$

Proof. For an arbitrary x^* we get

$$\begin{aligned}
(f_1 \oplus f_2)^*(x^*) &= \sup_x \{ \langle x^*, x\rangle - (f_1 \oplus f_2)(x) \} \\
&= \sup_x \Big\{ \langle x^*, x\rangle + \sup_{x_1+x_2=x} \{-f_1(x_1) - f_2(x_2)\} \Big\} \\
&= \sup_x \sup_{x_1+x_2=x} \{ \langle x^*, x_1\rangle - f_1(x_1) + \langle x^*, x_2\rangle - f_2(x_2) \} \\
&= \sup_{x_1,x_2} \{ \langle x^*, x_1\rangle - f_1(x_1) + \langle x^*, x_2\rangle - f_2(x_2) \} = f_1^*(x^*) + f_2^*(x^*).
\end{aligned}$$ □

Lemma 7.12. *For any functions $f_1, \dots, f_n$ we have*

$$(f_1 + \cdots + f_n)^* \le f_1^* \oplus \cdots \oplus f_n^*.$$

Proof. For an arbitrary x^* we have

$$\begin{aligned}
(f_1^* \oplus f_2^*)(x^*) &= \inf_{x_1^*+x_2^*=x^*} \{ f_1^*(x_1^*) + f_2^*(x_2^*) \} \\
&= \inf_{x_1^*+x_2^*=x^*} \{ \sup_{x_1} \{ \langle x_1^*, x_1\rangle - f_1(x_1) \} + \sup_{x_2} \{ \langle x_2^*, x_2\rangle - f_2(x_2) \} \} \\
&\ge \inf_{x_1^*+x_2^*=x^*} \{ \sup_x \{ \langle x_1^*, x\rangle - f_1(x) + \langle x_2^*, x\rangle - f_2(x) \} \} \\
&= \sup_x \{ \langle x^*, x\rangle - (f_1(x) + f_2(x)) \} = (f_1 + f_2)^*(x^*).
\end{aligned}$$ □

Theorem 7.13. *Let $f_1, \dots, f_n$ be convex proper functions such that all of them, with the exception of, possibly, one, are continuous at a certain point $x_0 \in \bigcap_{i=1}^n \operatorname{dom} f_i$. Then*

$$(f_1 + \cdots + f_n)^* = f_1^* \oplus \cdots \oplus f_n^*.$$

Proof. Basing on the previous lemma, it is sufficient to prove that $f_1^* \oplus f_2^* \le (f_1 + f_2)^*$. Let us do it. If $x^* \notin \operatorname{dom}(f_1 + f_2)^*$, then from the inequality $(f_1 + f_2)^* \le f_1^* \oplus f_2^*$ we get

$$(f_1 + f_2)^*(x^*) = (f_1^* \oplus f_2^*)(x^*) = +\infty.$$

Let $x^* \in \operatorname{dom}(f_1+f_2)^*$. Then $(f_1+f_2)^*(x^*) = \mu_0 < +\infty$. By the condition of the theorem, at least one of the functions, for determinacy, f_1 is continuous at the point $x_0 \in \operatorname{dom} f_1 \bigcap \operatorname{dom} f_2$. Then $x_0 \in \operatorname{int}(\operatorname{dom} f_1)$ and, hence, $x_0 \in \operatorname{int}(\operatorname{dom} f_1) \bigcap \operatorname{dom} f_2$. Therefore

$$\operatorname{dom}(f_1+f_2) \neq \varnothing \quad \text{and} \quad (f_1+f_2)^*(\xi^*) > -\infty \quad \forall \xi^* \in X.$$

In particular, $\mu_0 > -\infty$.

Consider the set

$$A = \{(x,\mu) \in X \times \mathbb{R} \colon \mu \leq \langle x^*, x\rangle - f_2(x) - \mu_0\}.$$

The set A is convex. Let us show that A does not intersect with the set

$$\operatorname{int}(\operatorname{epi} f_1) = \{(x,\mu) \in X \times \mathbb{R} \colon x \in \operatorname{int}(\operatorname{dom} f_1),\ \mu > f_1(x)\}.$$

If we suppose that there exists $(x,\mu) \in A \bigcap(\operatorname{int}(\operatorname{epi} f_1))$, then

$$\begin{aligned} &f_1(x) < \mu \leq \langle x^*, x\rangle - f_2(x) - \mu_0 \\ &\Rightarrow \mu_0 < \langle x^*, x\rangle - f_1(x) - f_2(x) \leq (f_1+f_2)^*(x^*) = \mu_0, \end{aligned}$$

that is impossible.

From here, by the theorem on separability there exist not simultaneously equal to zero $p \in X$ and $\alpha \in \mathbb{R}$, for which

$$\sup\{\langle p, x\rangle + \alpha\mu \colon (x,\mu) \in \operatorname{epi} f_1\} \leq \inf\{\langle p, x\rangle + \alpha\mu \colon (x,\mu) \in A\}. \tag{7.3}$$

Obviously, if $\alpha > 0$, then this inequality is violated. Hence $\alpha \leq 0$. If we assume $\alpha = 0$, then we obtain that the vector $p \neq 0$ separates the sets $\operatorname{dom} f_1$ and $\operatorname{dom} f_2$, contrary to condition $\operatorname{int}(\operatorname{dom} f_1) \bigcap \operatorname{dom} f_2 \neq \emptyset$. So, $\alpha < 0$. Without loss of generality we will assume that $\alpha = -1$.

Thus, by virtue of inequality (7.3) we have

$$\begin{aligned} f_1^*(p) &= \sup_{x\in X}\{\langle p, x\rangle - f_1(x)\} = \sup\{\langle p, x\rangle - \mu \colon (x,\mu) \in \operatorname{epi} f_1\} \\ &\leq \inf\{\langle p, x\rangle - \mu \colon (x,\mu) \in A\} = \inf\{\langle p - x^*, x\rangle + f_2(x) \colon x \in \operatorname{dom} f_2\} + \mu_0 \\ &= -f_2^*(x^* - p) + \mu_0. \end{aligned}$$

From this relation we get

$$(f_1^* \oplus f_2^*)(x^*) \leq f_1^*(p) + f_2^*(x^* - p) \leq \mu_0 = (f_1+f_2)^*(x^*),$$

concluding the proof of the theorem. □

8 Support functions

In various applications of convex analysis an important role is played by a convex function which is called *the support function of a set.*

Let X be a Euclidean space, and $A \subset X$ its nonempty subset.

Definition 8.1. The support function of a set $A \subset X$ is defined on X by the relation

$$c(x^*, A) = \sup_{y \in A} \langle x^*, y \rangle, \quad x^* \in X.$$

Present several simplest examples of the calculation of support functions.

Exercise 8.2. *For a given point $x \in X$ and a subset $A \subset X$ prove the formula*

$$c(x^*, A + x) = \langle x^*, x \rangle + c(x^*, A) \quad \forall x^* \in X.$$

Exercise 8.3. *Prove that the support function of the ball $A = O(x, r) = x + O(0, r)$ has the form*

$$c(x^*, O(x, r)) = \langle x^*, x \rangle + r|x^*|.$$

Exercise 8.4. *Let $A = \{x = (x_1, \dots, x_n) \in \mathbb{R}^n : |x_i| \le 1,\ i = 1, \dots, n\}$ be an n-dimensional cube. Prove that*

$$c(x^*, A) = |x_1^*| + \cdots + |x_n^*|, \quad \text{where } x^* = (x_1^*, \dots, x_n^*).$$

Consider certain properties of support functions.

Proposition 8.5. *Let $A \subset X$ be a nonempty set. Then:*

(1) *The support function is positively homogeneous in the first argument, i.e.,*

$$c(\lambda x^*, A) = \lambda c(x^*, A), \quad \forall \lambda \ge 0,\ \forall x^* \in X.$$

(2) *The support function is semi-additive in the first argument, i.e.,*

$$c(x_1^* + x_2^*, A) \le c(x_1^*, A) + c(x_2^*, A) \quad \forall x_1^*, x_2^* \in X.$$

(3) *The support function is convex and closed.*

(4) *The support function is additive in the second argument:*

$$c(\cdot, A + B) = c(\cdot, A) + c(\cdot, B).$$

(5) *Let $L: X \to X$ be a linear continuous operator. Then*

$$c(x^*, LA) = c(L^* x^*, A), \quad \forall x^* \in X.$$

(6) *If $A \subset B$, then*

$$c(x^*, A) \le c(x^*, B) \quad \forall x^* \in X.$$

(7) *Let $\delta_A(\cdot)$ be the indicator function of a convex closed set A. Then $c^*(\cdot, A) = \delta_A(\cdot)$.*

DOI 10.1515/9783110460308-009

Proof. (1) For $\lambda \geq 0$ we have

$$c(\lambda x^*, A) = \sup_{y\in A} \langle \lambda x^*, y\rangle = \sup_{y\in A} \lambda\langle x^*, y\rangle = \lambda \sup_{y\in A} \langle x^*, y\rangle = \lambda c(x^*, A).$$

$$(2)\ c(x_1^* + x_2^*, A) = \sup_{y\in A} \langle x_1^* + x_2^*, y\rangle = \sup_{y\in A} (\langle x_1^*, y\rangle + \langle x_2^*, y\rangle)$$

$$\leq \sup_{y\in A} \langle x_1^*, y\rangle + \sup_{y\in A} \langle x_2^*, y\rangle = c(x_1^*, A) + c(x_2^*, A).$$

(3) The convexity follows from the semi-additivity and positive homogeneity of the support function in the first argument. The closedness of the epigraph of a support function follows from the fact that it is the intersection of closed semi-spaces, which are the epigraphs of linear functions $\langle x^*, x\rangle$.

$$(4)\ c(x^*, A + B) = \sup_{y_1\in A,\, y_2\in B} \langle x^*, y_1 + y_2\rangle = \sup_{y_1\in A,\, y_2\in B} (\langle x^*, y_1\rangle + \langle x^*, y_2\rangle)$$

$$= \sup_{y_1\in A} \langle x^*, y_1\rangle + \sup_{y_2\in B} \langle x^*, y_2\rangle = c(x^*, A) + c(x^*, B).$$

(5) $c(x^*, LA) = \sup_{y\in LA} \langle x^*, y\rangle = \sup_{z\in A} \langle x^*, Lz\rangle = \sup_{z\in A} \langle L^* x^*, z\rangle = c(L^* x^*, A)$.

(6) For $A \subset B$ we have

$$c(x^*, A) = \sup_{y\in A} \langle x^*, y\rangle \leq \sup_{y\in B} \langle x^*, y\rangle = c(x^*, B).$$

(7) The function δ_A is convex, closed and proper. Therefore, by the Fenchel–Moreau theorem we have $\delta_A^{**} = \delta_A$. In addition, for an arbitrary x^* we get

$$\delta_A^*(x^*) = \sup_x \{\langle x, x^*\rangle - \delta_A(x)\} = \sup_{x\in A} \{\langle x, x^*\rangle\} = c(x^*, A).$$

Hence, $\delta_A = \delta_A^{**} = c^*(\cdot, A)$. □

Proposition 8.6.
(1) *The support function is proper.*
(2) *If $X = \mathbb{R}^n$ and a set A is bounded, then its support function is continuous.*
(3) *The support function is positively homogeneous in the second argument, i.e.,*

$$c(x^*, \lambda A) = \lambda c(x^*, A) \quad \forall \lambda \geq 0.$$

(4) *For each set A and B we have* $c(\cdot, A \cup B) = \max(c(\cdot, A), c(\cdot, B))$.
(5) *Let A, B be convex bounded subsets $\mathbb{R}^n$ and* $\operatorname{int} A \cap \operatorname{int} B \neq \emptyset$. *Then*

$$c(\cdot, A \cap B) = \operatorname{cl}(c(\cdot, A) \oplus c(\cdot, B)). \tag{8.1}$$

Proof. (1) Obviously, $c(x^*, A) > -\infty\ \forall x^* \in X$ and $c(0, A) = 0$, implying $c(\cdot, A) \not\equiv +\infty$.

(2) If a set is bounded, then its support function takes only finite values and, in the case of finite-dimensional X its convexity follows from its continuity.

(3) Follows from the definition of the support function.

(4) $c(x^*, A \cup B) = \sup_{y \in A \cup B} \langle x^*, y\rangle = \max\{\sup_{y\in A}\langle x^*, y\rangle, \sup_{y\in B}\langle x^*, y\rangle\} =$

$$= \max(c(x^*, A), c(x^*, B)).$$

(5) For a bounded set, the support function is convex and continuous on $\mathbb{R}^n$. Hence, by virtue of Lemma 7.11 and number (7) of Proposition 8.5 for an arbitrary x we have

$$(c(\cdot, A) \oplus c(\cdot, B))^*(x) = c^*(x, A) + c^*(x, B) = \delta_A(x) + \delta_B(x)$$
$$= \delta_{A\cap B}(x) = c^*(x, A \cap B),$$

implying, by virtue of Proposition 7.7:

$$(\mathrm{cl}\,\varphi)^* = c^*(\cdot, A \cap B), \tag{8.2}$$

where the function φ is defined by the relation

$$\varphi(x) = (c(\cdot, A) \oplus c(\cdot, B))(x).$$

Here we used the following property of indicator functions that can be easily verified: for each two sets A, B we get $\delta_A + \delta_B = \delta_{A\cap B}$. The function φ is proper, since it is not identically equal to $+\infty$, and, by virtue of that proved above, its conjugate function also does not identically equal $+\infty$. So, by Proposition 6.3 the function cl φ is also proper. Applying the Fenchel–Moreau theorem to equality (8.2), we get $c(\cdot, A \cap B) = \mathrm{cl}\,(c(\cdot, A) \oplus c(\cdot, B))$. □

Notice that assertion (5) of Proposition 8.6 may be simplified by omitting in (8.1) the operation of closure cl. This follows from the fact that in assumptions of number (5) the function φ is closed. The proof of this statement is found in [2], Theorem 16.4.

Assertion (2) of Proposition 8.6 may also be essentially strengthened. Namely, the following statement holds true.

Proposition 8.7. *Let a set $A \subset X$ be bounded and $|A| = \sup\{|x| : x \in A\}$. Then*

$$|c(x_1^*, A) - c(x_2^*, A)| \le |A||x_1^* - x_2^*| \quad \forall x_1^*, x_2^* \in X.$$

In other words, the support function $c(\cdot, A)$ satisfies the Lipschitz condition with the constant $|A|$.

Proof. By virtue of number (2) in Proposition 8.5 for arbitrary x_1^*, x_2^* we have

$$c(x_2^*, A) = c((x_2^* - x_1^*) + x_1^*, A) \le c(x_2^* - x_1^*, A) + c(x_1^*, A).$$

But by the definition of the support function and by virtue of the Cauchy–Bunyakovskii–Schwarz inequality $|c(x^*, A)| \le |x^*||A|\ \forall x^* \in X$. Hence $c(x_2^*, A) - c(x_1^*, A) \le |x_2^* - x_1^*||A|$. Interchanging x_1^* and x_2^*, finally we obtain $|c(x_1^*, A) - c(x_2^*, A)| \le |A|\,|x_1^* - x_2^*|$. □

Proposition 8.8. *For each set $A \subset X$ we have*

$$c(\cdot, A) = c(\cdot, \operatorname{conv} A).$$

Proof. Since $A \subset \operatorname{conv} A$, we get $c(\cdot, A) \le c(\cdot, \operatorname{conv} A)$. Prove the inverse inequality. Indeed, assume that it is violated. Then there exists x^* such that $c(x^*, A) < c(x^*, \operatorname{conv} A)$. Therefore, there exists $y \in \operatorname{conv} A$ for which $c(x^*, A) < \langle x^*, y\rangle$, i.e., $\langle x^*, z\rangle < \langle x^*, y\rangle$ $\forall z \in A$.

By Theorem 2.2, conv A consists of all possible convex combinations of elements from A. Therefore there exist a number k, vectors $z_1, \dots z_k \in A$ and numbers $\lambda_1 > 0, \dots, \lambda_k > 0$ such that

$$\sum_{i=1}^{k} \lambda_i = 1, \quad y = \lambda_1 z_1 + \dots + \lambda_k z_k$$

and, hence,

$$\langle x^*, y\rangle = \sum_{i=1}^{k} \lambda_i \langle x^*, z_i\rangle < \sum_{i=1}^{k} \lambda_i \langle x^*, y\rangle = \langle x^*, y\rangle.$$

The obtained contradiction completes the proof. □

Proposition 8.9. *Let a set $A \subset X$ be convex and closed. Then*

$$A = \{x \in X \colon \langle x^*, x\rangle \le c(x^*, A)\ \forall x^* \in X\}.$$

Proof. The inclusion

$$A \subseteq \{x \in X \colon \langle x^*, x\rangle \le c(x^*, A)\ \forall x^* \in X\}$$

is evident. Prove the inverse inclusion, i.e., if $\langle x^*, x\rangle \le c(x^*, A)$ $\forall x^* \in X$, then $x \in A$. Indeed, suppose that $x \notin A$. Then the point x may be strictly separated from the convex closed set A and, hence, there exist such $x^* \in X$ and $\alpha \in \mathbb{R}$, that

$$\langle x^*, x\rangle > \alpha \ge \langle x^*, \xi\rangle\ \forall \xi \in A \Rightarrow \langle x^*, x\rangle > c(x^*, A).$$

The obtained contradiction completes the proof. □

The proved assertion may be represented in the following equivalent form

$$x \in A \Leftrightarrow \langle x^*, x\rangle \le c(x^*, A) \quad \forall x^* \in X. \tag{8.3}$$

Proposition 8.10. *Let A, B be convex closed sets and*

$$c(x^*, A) \le c(x^*, B) \quad \forall x^* \in X.$$

Then $A \subset B$.

Proof. Let $x \in A$. Then, by virtue of the previous assertion, $\langle x^*, x\rangle \le c(x^*, A) \le c(x^*, B)$ $\forall x^*$ and, hence, $x \in B$. □

Proposition 8.11. *Let A, B be convex closed sets and*

$$c(\cdot, A) = c(\cdot, B).$$

Then $A = B$.

Proof. Since $c(\cdot, A) = c(\cdot, B)$, we have $c(\cdot, A) \le c(\cdot, B)$ and $c(\cdot, B) \le c(\cdot, A)$, yielding, by virtue of Proposition 8.10: $A \subset B$ and $B \subset A$ and, hence, $A = B$. □

Proposition 8.12. *Let A and B convex closed sets. Then*

$$A \bigcap B \ne \varnothing \Longleftrightarrow c(x^*, A) + c(-x^*, B) \ge 0 \quad \forall x^* \in X.$$

Proof. Let $A \bigcap B \ne \varnothing$. Then

$$\begin{aligned} 0 \in A + (-B) \Rightarrow 0 &\le c(x^*, A + (-B)) \\ &= c(x^*, A) + c(x^*, -B) = c(x^*, A) + c(-x^*, B) \quad \forall x^* \in X. \end{aligned}$$

Prove the inverse implication. For an arbitrary x^* we have

$$0 \le c(x^*, A) + c(-x^*, B) = c(x^*, A) + c(x^*, -B) = c(x^*, A - B),$$

implying, by virtue of (8.3), $0 \in A - B$ and, hence, $A \bigcap B \ne \varnothing$. □

Proposition 8.13. *Suppose that a set A is convex and* $\operatorname{int} A \ne \emptyset$. *Then*

$$\langle x^*, x\rangle < c(x^*, A)\ \forall x^* \ne 0 \Leftrightarrow x \in \operatorname{int} A.$$

Proof. Indeed, let $x \in \operatorname{int} A$. Then there exists $\varepsilon > 0$ such that $O(x, \varepsilon) \subset A$. Therefore, for an arbitrary $x^* \ne 0$ we have

$$c(x^*, O(x, \varepsilon)) \le c(x^*, A) \Rightarrow \langle x^*, x\rangle + \varepsilon|x^*| \le c(x^*, A),$$

yielding $\langle x^*, x\rangle < c(x^*, A)$.

Suppose now that $x \notin \operatorname{int} A$. Then, since $\operatorname{int} A \ne \emptyset$, by the separability theorem the point x may be separated from the convex set $\operatorname{int} A$. Then there exists $x^* \ne 0$ such that $\langle x^*, x\rangle \ge \langle x^*, \xi\rangle\ \forall \xi \in A$. Hence, $\langle x^*, x\rangle \ge c(x^*, A)$, proving the desired. □

As we demonstrated above, the support function is convex, closed, positively homogeneous and proper. There arises the natural question: if a certain function obeys all the above properties, should it be a support function for a certain set? The following assertion gives an affirmative answer to this question.

Theorem 8.14. *A function f is the support function for certain nonempty convex and closed sets if and only if it is a convex, closed, positively homogeneous, and proper function.*

Proof. The *necessity* was justified above.

Sufficiency. Set $g = f^*$. Then g is a convex, closed and proper function. Hence, by the Fenchel–Moreau theorem, $g^* = f^{**} = f$. Let us prove that g may take only two values: 0 and $+\infty$. In fact, for an arbitrary $\lambda > 0$ we have

$$\begin{aligned}(\lambda g)^*(x^*) &= \sup_x(\langle x^*, x\rangle - \lambda g(x)) \\ &= \lambda \sup_x(\langle x^*/\lambda, x\rangle - g(x)) = \lambda g^*(x^*/\lambda) = \lambda f(x^*/\lambda) \\ &= f(x^*) = g^*(x^*) \quad \Rightarrow (\lambda g)^* = g^*.\end{aligned}$$

Therefore, by the Fenchel–Moreau theorem, $\lambda g = g$ and, hence, the function g may take only two values: 0 and $+\infty$.

Consider the convex set $A = \{x\colon g(x) \le 0\}$. Then, as has been proved, $A = \{x\colon g(x) = 0\}$, and $A \ne \emptyset$, since by Lemma 7.9 the function g is proper. Obviously, also $g = \delta_A$, implying $g^* = \delta_A^* = c(\cdot, A)$ and hence $f = g^* = c(\cdot, A)$. □

Corollary 8.15. *Let f be a positively homogeneous, convex, proper function and there exist $a \in X$ and $b \in \mathbb{R}$ such that*

$$f(x) \ge \langle a, x\rangle + b \quad \forall x \in X. \tag{8.4}$$

Then

$$\operatorname{cl} f = c(\cdot, A), \quad \textit{where } A = \{x\colon \langle x^*, x\rangle \le f(x^*)\ \forall x^* \in X\}. \tag{8.5}$$

Proof. It is easy to see that the epigraph of a convex positively homogeneous function is a convex cone (i.e., convex set, containing with each of its points x also the ray radiated by it: $\cup_{\lambda>0}\lambda x$). Therefore the function $\operatorname{cl} f$ is convex, closed and positively homogeneous. In addition, $\operatorname{cl} f$ is a proper function, since $\operatorname{cl} f \le f$ and by virtue of (8.4), $(\operatorname{cl} f)(x) \ne -\infty\ \forall x \in X$.

By the proved theorem, there exists such a convex closed set A, that $\operatorname{cl} f = c(\cdot, A)$. Using number (7) of Proposition 8.5 and the fact that, by Proposition 7.7 $(\operatorname{cl} f)^* = f^*$, we get $f^* = (\operatorname{cl} f)^* = c^*(\cdot, A) = \delta_A$. Therefore

$$A = \{x\colon \delta_A(x) \le 0\} = \{x\colon f^*(x) \le 0\} = \{x\colon \langle x^*, x\rangle \le f(x^*)\ \forall x^*\}.$$ □

Corollary 8.16. *Let f be a positively homogeneous, convex, proper function and $X = \mathbb{R}^n$. Then* (8.5) *holds true.*

Proof. The validity of this statement follows immediately from the previous corollary and Lemma 5.4. □

So, each convex closed set determines its support function, which belongs to the class of functions $\mathcal{F}$, consisting of all convex, closed, positively homogeneous, and proper functions. By virtue of the obtained results, the inverse is also true, namely: each function f belonging to the class $\mathcal{F}$ is the support function for some set A, from the collection of subsets $\mathcal{A}$, consisting of all convex, closed subsets of X and, moreover, this set A is determined uniquely, by virtue of (8.5).

Therefore, there exists a one-to-one correspondence between convex closed sets and the functions of class $\mathcal{F}$, and, moreover, this correspondence is described by relations (8.5). This correspondence presents the example of duality between two classes of objects of different nature: functions $f \in \mathcal{F}$ and sets $A \in \mathcal{A}$. During the investigation of convex closed sets, such duality allows us to use the theory of functions and, conversely, to use the results of the theory of convex sets (e.g., the separability theory etc.) for studying the functions from the class $\mathcal{F}$. Instead of convex closed sets, we can also use their support functions which form the class $\mathcal{F}$.

9 Differentiability of convex functions and the subdifferential

Everywhere in this section (with the exception of Lemma 9.13) we will assume that $X = \mathbb{R}^n$.

It turns out that besides the continuity on the interior of its domain, a proper convex function on $\mathbb{R}^n$ has some additional properties. Namely, by virtue of Theorem 6.12 a convex function f is also locally Lipschitz on $\operatorname{int}(\operatorname{dom} f)$, i.e., each point $x \in \operatorname{int}(\operatorname{dom} f)$ has such a neighborhood O on which f satisfies the Lipschitz condition. But, by the Rademacher[1] theorem (see [33]) each locally Lipschitz function on $\mathbb{R}^n$ is differentiable almost everywhere, i.e., the set of points in which this function is not differentiable has zero Lebesgue measure[2]. Hence, a proper convex function is differentiable almost everywhere on the set $\operatorname{int}(\operatorname{dom} f)$.

Let us study differential properties of a proper convex function f and consider the analogue of its derivative, which is called *the subdifferential of a function f* at a given point $x \in \operatorname{int}\operatorname{dom} f$.

Definition 9.1. Suppose that a function f has a finite value at a point x. The derivative of a function f at a point x along a direction y is the limit

$$f'(x; y) = \lim_{\lambda \to 0+} \frac{f(x + \lambda y) - f(x)}{\lambda},$$

if it exists (the infinite values of this limit are admissible).

If a function f is differentiable at a point x then, as it is known, $f'(x; y) = \langle \operatorname{grad} f(x), y \rangle$, where the coordinates of the gradient $\operatorname{grad} f(x)$ are the partial derivatives of the function f at the point x.

Theorem 9.2. *Let a function f be convex and finite at a point x. Then*

$$f'(x; y) = \inf_{\lambda > 0} \frac{f(x + \lambda y) - f(x)}{\lambda}.$$

Moreover, the function $f'(x; \cdot)$ is convex, positively homogeneous and

$$f'(x; 0) = 0; \quad -f'(x; -y) \leq f'(x; y) \ \forall y.$$

Proof. For convenience, let us assume that $x = 0$, $f(x) = 0$. For $\lambda > 0$, set $h(\lambda) = \frac{f(\lambda y)}{\lambda}$. Let us show that the function $h(\cdot)$ is not decreasing. Indeed, let $0 < \mu < \lambda$. Then, by

1 Hans Adolph Rademacher (1892–1969), a German mathematician.

2 Necessary information on measure theory may be found, for example, in [27].

DOI 10.1515/9783110460308-010

virtue of convexity of f we have

$$f(\mu y) = f\left(0\left(1-\frac{\mu}{\lambda}\right) + \frac{\mu}{\lambda}\lambda y\right) \le f(0)\left(1-\frac{\mu}{\lambda}\right) + f(\lambda y)\frac{\mu}{\lambda} = \frac{\mu}{\lambda} f(\lambda y)$$
$$\Rightarrow \frac{f(\mu y)}{\mu} \le \frac{f(\lambda y)}{\lambda}$$

and hence h is not decreasing. Therefore,

$$\lim_{\lambda\to 0+} h(\lambda) = \inf_{\lambda>0} h(\lambda) = f'(0;y).$$

Let $\mu \ge 0$ be given. Then obviously $\frac{f(\lambda(\mu y))}{\lambda} = \mu\frac{f(\kappa y)}{\kappa} = \mu h(\kappa)$, where $\kappa = \lambda\mu$. So, $f'(0;(\mu y)) = \mu \lim_{\kappa\to 0+} h(\kappa) = \mu f'(0;y)$ and, hence, function $f'(0;\cdot)$ is positively homogeneous and $f'(0;0) = 0$ by definition.

Let us prove the convexity of the function $f'(x;\cdot)$. Let $\alpha \ge 0$, $\beta \ge 0$, $\alpha+\beta = 1$, $y_1, y_2 \in X$. By the convexity of f we have

$$\begin{aligned} f'(x;\alpha y_1 + \beta y_2) &= \lim_{\lambda\to 0+} \frac{f(x+\lambda\alpha y_1 + \lambda\beta y_2) - f(x)}{\lambda} \\ &\le \lim_{\lambda\to 0+} \frac{\alpha f(x+\lambda y_1) + \beta f(x+\lambda y_2) - \alpha f(x) - \beta f(x)}{\lambda} \\ &= \lim_{\lambda\to 0+} \alpha\frac{f(x+\lambda y_1) - f(x)}{\lambda} + \beta \lim_{\lambda\to 0+} \frac{f(x+\lambda y_2) - f(x)}{\lambda} \\ &= \alpha f'(x;y_1) + \beta f'(x;y_2). \end{aligned}$$

Hence the function $f'(x;\cdot)$ is convex. From its convexity and positive homogeneity we get

$$f'(x;y) + f'(x;-y) \ge 2f'(x;\frac{y+(-y)}{2}) = 2f'(x;0) = 0,$$

yielding $-f'(x;-y) \le f'(x;y)$. □

Lemma 9.3. *Let f be a proper convex function and $x \in \operatorname{int}(\operatorname{dom} f)$. Then there exists $c > 0$ such that*

$$|f'(x;y)| \le c|y| \quad \forall y.$$

Proof. By Theorem 6.12 there exists such a neighborhood O of the point x that the function f satisfies on O the Lipschitz condition with a certain constant $c > 0$. Then, for each fixed y, the inequality $|(f(x+\lambda y) - f(x))/\lambda| \le c|y|$ holds true for all sufficiently small $\lambda > 0$. The desired statement immediately follows from this inequality and the relation

$$f'(x;y) = \lim_{\lambda\to 0+} \frac{f(x+\lambda y) - f(x)}{\lambda}.$$ □

Let us introduce one of the most important notions of convex analysis.

Definition 9.4. A vector $x^* \in X$ is called the subgradient of a function f at a point x if

$$f(y) \ge f(x) + \langle x^*, y - x\rangle \quad \forall y \in X.$$

This relation is said to be *the subgradient inequality.*

Definition 9.5. The set of all subgradients of a function f at a point x is called the subdifferential of f at this point. The subdifferential is denoted by $\partial f(x)$.

Let us make an evident remark: a function f achieves the minimum at a point x if and only if $0 \in \partial f(x)$.

Exercise 9.6. *Prove that if $f(x) = |x|$ then $\partial f(0) = [-1, +1]$.*

Decision. If $x^* \in \partial f(0)$ then $f(y) = |y| \geq \langle x^*, y\rangle$ $\forall y \in \mathbb{X}$, hence $|x^*| \leq 1$. Conversely, if $|x^*| \leq 1$, then by the Cauchy–Bunyakovskii–Schwarz inequality $f(y) = |y| \geq \langle x^*, y\rangle$ $\forall y \in X$, yielding $x^* \in \partial f(0)$.

Lemma 9.7. *Let a function f be convex and finite at a point x. Then*

$$x^* \in \partial f(x) \Leftrightarrow f'(x; y) \geq \langle x^*, y\rangle \quad \forall y \in X.$$

Proof. Let $x^* \in \partial f(x)$. Then, by the subgradient inequality, for each $\lambda > 0$ we have $f(x + \lambda y) - f(x) \geq \langle x^*, \lambda y\rangle$ $\forall y$ and hence

$$f'(x; y) = \lim_{\lambda \to 0+} \frac{f(x + \lambda y) - f(x)}{\lambda} \geq \langle x^*, y\rangle .$$

Now, conversely, let

$$\langle x^*, y\rangle \leq f'(x; y) = \inf_{\lambda > 0} \frac{f(x + \lambda y) - f(x)}{\lambda} \quad \forall y$$
$$\Rightarrow f(x + \lambda y) - f(x) \geq \langle x^*, \lambda y\rangle \quad \forall y \in X,\ \forall \lambda > 0.$$

Therefore, for $\lambda = 1$, $z = x + y$ we have $f(z) - f(x) \geq \langle x^*, z - x\rangle$ $\forall z \in X$, yielding $x^* \in \partial f(x)$. □

Lemma 9.8. *Let a function f be convex, proper and $x \in \operatorname{int}(\operatorname{dom} f)$. Then the function $\operatorname{cl} f'(x; \cdot)$, which is the closure of the derivative $f'(x; y)$ as the convex function on y, coincides with the support function of the set $\partial f(x)$.*

Proof. By Theorem 9.2 and Lemma 9.3, the function $f'(x; \cdot)$ is convex, positively homogeneous and proper. So, by virtue of Corollary 8.16 we get $\operatorname{cl} f'(x; \cdot) = c(\cdot, A)$, where $A = \{x^* : \langle x^*, y\rangle \leq f'(x; y)\ \forall y\}$. But by Lemma 9.7 we have $A = \partial f(x)$. □

Let us describe the main properties of the subdifferential. From the subgradient inequality it immediately follows that if f is a proper function and $x \notin \operatorname{dom} f$, then $\partial f(x) = \emptyset$.

Theorem 9.9. *Let f be a convex proper function and $x \in \operatorname{int}(\operatorname{dom} f)$. Then the subdifferential $\partial f(x)$ at a point x is a nonempty convex compact set.*

Proof. First, prove that $\partial f(x) \neq \emptyset$. Consider the set $\operatorname{epi} f$. It is convex and $(x, f(x)) \notin \operatorname{int}(\operatorname{epi} f) \neq \emptyset$. The last assertion easily follows from the fact that a convex proper function defined on $\mathbb{R}^n$ is continuous on int (dom f). Therefore, by Theorem 4.7 on separability, there exist $x^* \in \mathbb{R}^n$, $r \in \mathbb{R}$ such that

$$r\alpha + \langle x^*, z\rangle \geq rf(x) + \langle x^*, x\rangle, \quad \forall \alpha, z\colon \alpha \geq f(z).$$

Condition $r \neq 0$ follows from $x \in \text{int}(\text{dom} f)$. Moreover, $r > 0$ since the presented inequality holds true for arbitrarily large $\alpha \geq f(z)$. Hence, without loss of generality, we will assume that $r = 1$. From the same inequality for $\alpha = f(z)$ we have

$$f(z) \geq f(x) + \langle -x^*, z - x\rangle \ \forall z \Rightarrow -x^* \in \partial f(x)$$

and so $\partial f(x) \neq \emptyset$.

Let us show the boundedness of the set $\partial f(x)$. In fact, suppose the contrary, i.e., that in $\partial f(x)$ exists such a sequence $\{x_i^*\}$, that $|x_i^*| \to \infty$. Choose such $\delta > 0$, that $\text{cl}(O(x, \delta)) \subset \text{int}(\text{dom} f)$. Then the function f is continuous and hence bounded on the compact set $\text{cl}(O(x, \delta))$. Take $x_i = x + \delta x_i^*/|x_i^*|$. From the subgradient inequality for $y = x_i$ and $x^* = x_i^*$ we get $f(x_i) \geq f(x) + \delta|x_i^*|$. But in the obtained inequality the right-hand side tends to infinity whereas the left-hand side is bounded since $x_i \in \text{cl}(O(x, \delta))\ \forall i$. This contradiction shows the boundedness of the set $\partial f(x)$.

Let us prove that the set $\partial f(x)$ is closed. Let $x_i^* \in \partial f(x)\ \forall i$ and $x_i^* \to x^*$. For each fixed y substituting $x^* = x_i^*$ into the subgradient inequality and, passing to the limit as $i \to \infty$, we get $x^* \in \partial f(x)$.

It remains to show the convexity of the set $\partial f(x)$. Let $x_1^*, x_2^* \in \partial f(x)$, $\alpha \geq 0$, $\beta \geq 0$, $\alpha + \beta = 1$. Then

$$f(z) - f(x) \geq \langle x_1^*, z - x\rangle, \quad f(z) - f(x) \geq \langle x_2^*, z - x\rangle \quad \forall z \in X.$$

For a fixed z, multiply the first of these inequalities by α, and the second one by β, and then add them. As the result, we get $\alpha x_1^* + \beta x_2^* \in \partial f(x)$. □

Theorem 9.9 is proved for the case $X = \mathbb{R}^n$. Let us present its infinite-dimensional analogue. First we mention that the notions of subgradient, subgradient inequality and subdifferential preserve the same meaning for an arbitrary (possibly, infinite-dimensional) normed space X, with the only difference that x^* is the element of the space X^*, conjugate to X, consisting of linear continuous functionals on X, and the notation $\langle x^*, x\rangle$ means the action of the linear functional x^* on the vector $x \in X$.

Theorem 9.10. *Let X be an infinite-dimensional normed space, f a convex proper function on X, bounded above on a certain neighborhood of a point $x \in \text{int}(\text{dom} f)$. Then the subdifferential $\partial f(x)$ at a point x is a nonempty, convex, bounded and closed subset of X^*.*

The proof of this assertion follows the same lines as the proof of Theorem 9.9. The only difference consists in the use of Theorem 6.8 on continuity.

Let us return to the finite-dimensional case. If $x \notin \text{dom} f$ then, by virtue of the gradient inequality we have $\partial f(x) = \emptyset$. The following example shows that if x belongs not to the interior of the set dom f, but only to its boundary, then the subdifferential $\partial f(x)$ also may be empty.

Exercise 9.11. *Consider on* $\mathbb{R}$ *the function*

$$f(x) = \begin{cases} -(1-x^2)^{1/2}, & |x| \le 1, \\ +\infty, & |x| > 1. \end{cases}$$

This function is subdifferentiable (and even differentiable) in all points $x \in \operatorname{int}(\operatorname{dom} f) = (-1, 1)$. *Prove that, nevertheless, at points* $x = -1, 1$, *forming the boundary of* dom f *its subdifferential is empty.*

This situation may be slightly refined. Namely, the following fairly unusual assertion holds.

Lemma 9.12 (The finite-dimensional alternative). *Let a convex function f be finite at a point x_0, belonging to the boundary of the domain* dom f. *Then the subdifferential* $\partial f(x_0)$ *at this point is either empty or contains infinitely many points.*

Proof. By virtue of convexity of the subdifferential $\partial f(x_0)$ it is sufficient to prove that if this set is nonempty, then it contains at least two distinct elements. So, let $x^* \in \partial f(x_0)$. Consider the convex function $h(y) = f(x_0 + y) - f(x_0) - \langle x^*, y\rangle$. We have $h(0) = 0$ and zero belongs to the boundary of the domain dom h. In addition, $0 \in \partial h(0)$, from which, by the subgradient inequality we get $h(y) \ge 0\ \forall y \in X$.

From the above mentioned it follows that $0 \notin \operatorname{int}(\operatorname{dom} h)$. Hence, the convex set dom h may be separated from zero and so, by Theorem 4.6 on the finite-dimensional separability there exists such an $a \in X$, $a \ne 0$, that $\langle a, y\rangle \le 0\ \forall y \in \operatorname{dom} h$. Therefore, $h(y) \ge \langle a, y\rangle\ \forall y \in X$, since $h(y) \ge 0\ \forall y \in X$. It means that $0, a \in \partial h(0)$, $a \ne 0 \Rightarrow x^*, (a + x^*) \in \partial f(x_0)$. □

Lemma 9.12 also has an infinite-dimensional analogue.

Lemma 9.13 (The infinite-dimensional alternative). *Let X be an arbitrary normed space and a convex function f on X is finite at a point x_0, belonging to the boundary of the domain* dom f *and, moreover, the interior of the domain* int (dom f) *is nonempty. Then the subdifferential* $\partial f(x_0)$ *at this point is either empty or contains infinitely many points.*

The proof of this assertion differs from the proof of Lemma 9.12 in the following: instead of Theorem 4.6 on finite-dimensional separability one needs to use the separability Theorem 4.7.

Now, let us consider the following question: when is the subdifferential of the sum of functions equal to the sum of their subdifferentials?

Proposition 9.14. *For any functions $f_1, \dots, f_k$ we have*

$$\partial(f_1 + \dots + f_k)(x) \supseteq \partial f_1(x) + \dots + \partial f_k(x), \quad \forall x \in X.$$

Proof. Let $x^* \in \partial f_1(x) + \dots + \partial f_k(x)$. Then $x^* = x_1^* + \dots + x_k^*$, where $x_i^* \in \partial f_i(x)$. Therefore

$$f_1(y) \geq f_1(x) + \langle x_1^*, y - x \rangle, \dots, f_k(y) \geq f_k(x) + \langle x_k^*, y - x \rangle \quad \forall y \in X.$$

Adding, for each fixed y, all these inequalities we get $x^* \in \partial(f_1 + \dots + f_k)(x)$. □

It turns out that the converse inequality is true only for additional, although rather weak assumptions.

Theorem 9.15 (Moreau–Rockafellar[3]). *Let functions $f_1, \dots, f_k$ be proper, convex and there exist a point $x_0 \in \bigcap_i \operatorname{dom} f_i$, in which all functions f_i, except, possibly, one, are continuous. Then*

$$\partial(f_1 + \dots + f_k)(x) = \partial f_1(x) + \dots + \partial f_k(x) \quad \forall x \in X.$$

Proof. Let us give the proof only for the case $k = 2$ (the general case may be verified by induction). From the previous assertion, it follows that it is sufficient to show that

$$\partial(f_1 + f_2)(x) \subseteq \partial f_1(x) + \partial f_2(x) \quad \forall x \in X. \tag{9.1}$$

By condition of the theorem, at least one of the functions, say f_1, is continuous at a point $x_0 \in \operatorname{dom} f_1 \bigcap \operatorname{dom} f_2$. Then $x_0 \in \operatorname{int}(\operatorname{dom} f_1)$ and hence the intersection of the sets $\operatorname{int}(\operatorname{dom} f_1)$ and $\operatorname{dom} f_2$ is nonempty.

Fix $x \in X$. Take arbitrary $x^* \in \partial(f_1 + f_2)(x)$. Replacing, if necessary, the functions f_1, f_2 with the proper convex functions

$$g_1(z) = f_1(x + z) - f_1(x) - \langle x^*, z \rangle, \quad g_2(z) = f_2(x + z) - f_2(x),$$

we will assume, without loss of generality, that

$$x = 0, \quad f_1(0) = 0, \quad f_2(0) = 0, \quad x^* = 0.$$

So, $0 \in \partial(f_1 + f_2)(0)$. Let us prove that $0 \in \partial f_1(0) + \partial f_2(0)$. Indeed, by the assumption, we have

$$f_1(z) + f_2(z) \geq f_1(0) + f_2(0) = 0 \quad \forall z \in X. \tag{9.2}$$

Consider the sets

$$C_1 = \{(z, \mu) : \mu \geq f_1(z)\}, \quad C_2 = \{(z, \mu) : \mu < -f_2(z)\}.$$

From the convexity of functions f_1, f_2 immediately follows that both these sets are convex. Moreover, they are disjoint since otherwise we should have the inequality

3 Ralph Tyrrell Rockafellar (born 1935), an American mathematician.

$-f_2(z) > f_1(z)$ at a certain point z, contrary to (9.2). So, by the finite-dimensional separability theorem these two sets may be separated, i.e., there exist such $z^* \in \mathbb{R}^n$ and $\beta \in \mathbb{R}$, not equal to zero simultaneously, that

$$\sup_{(z,\mu)\in C_1} (\beta\mu + \langle z^*, z\rangle) \leq \inf_{(z,\mu)\in C_2} (\beta\mu + \langle z^*, z\rangle). \tag{9.3}$$

It is easy to see that $\beta \leq 0$, since for $\beta > 0$ the upper bound in (9.3) equals $+\infty$ and lower lower bound is $-\infty$. The case $\beta = 0$ is also impossible since if $\beta = 0$, then $z^* \neq 0$ and (9.3) takes the form

$$\sup_{z\in \operatorname{dom} f_1} \langle z^*, z\rangle \leq \inf_{z\in \operatorname{dom} f_2} \langle z^*, z\rangle.$$

This is in the contradiction of the fact that the intersection of the sets $\operatorname{int}(\operatorname{dom} f_1)$ and $\operatorname{dom} f_2$ is nonempty. Therefore, it is proved that $\beta < 0$ and so, without loss of generality we will assume that $\beta = -1$.

From (9.3) the next inequality follows:

$$\sup_z\{\langle z^*, z\rangle - f_1(z)\} \leq \inf_z\{\langle z^*, z\rangle + f_2(z)\}.$$

But for $z = 0$ the expressions in braces in both sides of this inequality vanish. Therefore we get

$$f_1(z) \geq \langle z^*, z\rangle, \quad f_2(z) \geq \langle -z^*, z\rangle \quad \forall z \in X.$$

So,

$$f_1(z) \geq f_1(0) + \langle z^*, z - 0\rangle \qquad \forall z \in X,$$
$$f_2(z) \geq f_2(0) + \langle -z^*, z - 0\rangle \qquad \forall z \in X.$$

Hence $z^* \in \partial f_1(0)$, $-z^* \in \partial f_2(0)$ yielding $0 \in \partial f_1(0) + \partial f_2(0)$. □

Let us give the example which demonstrates that inclusion (9.1) may be violated without an additional supposition concerning the nonemptiness of the intersection of interiors of domains of convex functions.

Exercise 9.16. *Consider the following convex functions on* $\mathbb{R}$*:*

$$f_1(x) = \begin{cases} -\sqrt{x}, & x \geq 0 \\ +\infty, & x < 0 \end{cases}; \qquad f_2(x) = \begin{cases} -\sqrt{-x}, & x \leq 0 \\ +\infty, & x > 0 \end{cases}.$$

Take a point $x = 0$*. Obviously,* $(f_1 + f_2)(x) = +\infty$, $x \neq 0$, $(f_1 + f_2)(0) = 0$*. Hence* $\partial(f_1 + f_2)(0) = (-\infty, +\infty)$*.*

At the same time, prove that $\partial f_1(0) = \partial f_2(0) = \varnothing$ *and hence* $\partial f_1(0) + \partial f_2(0) = \varnothing$*. Therefore for* $x = 0$ *inclusion (9.1) is violated.*

Let us consider the question about the connection of the structure of the subdifferential of a convex function with its differentiability. We will show that the differentiability of a convex function at a certain point is equivalent to the fact that its subdifferential at this point consists of a single element.

Theorem 9.17. *Let a convex function f be finite at a point x_0. If f is differentiable at a point x_0, then the subdifferential $\partial f(x_0)$ contains an unique element $f'(x_0)$ and, in particular,*

$$f(z) \geq f(x_0) + \langle f'(x_0), z - x_0 \rangle \quad \forall z \in X.$$

Conversely, if the subdifferential $\partial f(x_0)$ consists of a single element, then the function f is differentiable at the point x_0 and $\partial f(x_0) = \{ f'(x_0) \}$.

Proof. Let f be differentiable at a point x_0. Then $f'(x_0; y) = \langle f'(x_0), y \rangle \ \forall y$. Hence, by virtue of Lemma 9.7 for each $x^* \in \partial f(x_0)$ we have

$$\langle f'(x_0), y \rangle = f'(x_0; y) \geq \langle x^*, y \rangle \quad \forall y \in X,$$

yielding

$$\langle (f'(x_0) - x^*), y \rangle \geq 0 \quad \forall y \in X.$$

But if a linear functional is nonnegative on the whole space, then it is equal to zero and hence $x^* = f'(x_0) \ \forall x^* \in \partial f(x_0)$, i.e., the subdifferential $\partial f(x_0)$ consists of a single point $f'(x_0)$.

Now, let $\partial f(x_0) = \{x^*\}$. Consider the convex function $h(y) = f(x_0 + y) - f(x_0) - \langle x^*, y \rangle$. We have $h(0) = 0$, $\partial h(0) = \{0\}$, from which, by the subgradient inequality, we get $h(y) \geq 0 \ \forall y \in X$ and hence the function h is proper. In addition, by virtue of Lemma 9.12 we have $0 \in \operatorname{int}(\operatorname{dom} h)$. Therefore, by Lemma 9.8, the function $\operatorname{cl} h'(0; \cdot)$ is the support function for the set $\partial h(0)$. Hence, $\operatorname{cl} h'(0; y) = c(y, \partial h(0)) = 0 \ \forall y \in X$.

So $0 \in \operatorname{int}(\operatorname{dom} h)$, and h is the proper convex function. By Lemma 9.3 the convex function $h'(0; \cdot)$ takes only finite values and, hence it is continuous on $\mathbb{R}^n$. Therefore by that proved above, $h'(0; y) \equiv 0$. This yields

$$\lim_{\lambda \to 0+} \frac{h(\lambda y)}{\lambda} = 0 \quad \forall y \in X.$$

Besides, from the proof of Theorem 9.2 it follows that for each fixed y, the function $g(\lambda) = h(\lambda y)/\lambda$ is nondecreasing for $\lambda > 0$.

For each $\lambda > 0$, consider the function $g_\lambda(u) = h(\lambda u)/\lambda$. Each of the functions g_λ is convex and for each u, by virtue of that proved above, we have $g_\lambda(u) \to 0$, $\lambda \to 0+$, $g_\lambda(u) \geq 0$. Let $B = \{ x \colon |x| \leq 1 \}$ be the unit ball and $\{ b_1, \dots, b_m \}$ a finite collection of points whose convex hull contains B.

Each $u \in B$ may be represented in the form of a linear combination

$$u = \alpha_1 b_1 + \cdots + \alpha_m b_m.$$

Therefore, since each of the functions g_λ is convex, for each $u \in B$ we have

$$0 \leq g_\lambda(u) \leq \alpha_1 g_\lambda(b_1) + \cdots + \alpha_m g_\lambda(b_m) \leq \max\{ g_\lambda(b_i), \ i = 1, \dots, m \}.$$

Now, since $g_\lambda(b_i) \to 0$, $\lambda \to 0+$ for each i, then the functions g_λ tend to zero uniformly on B. So, for each $\varepsilon > 0$ there exists $\delta > 0$ such that

$$h(\lambda u)/\lambda \leq \varepsilon \quad \forall \lambda \in (0, \delta], \ \forall u \in B.$$

But each vector $y \neq 0$, $|y| \leq \delta$ may be represented in the form $y = \lambda u$, where $\lambda = |y| \leq \delta$, $u = y/|y| \in B$. Therefore the inequality $h(y)/|y| \leq \varepsilon$ holds for all y, for which $|y| \leq \delta$.

So we proved that $h(y)/|y| \to 0$, $y \to 0$. In other words, the function h is differentiable at zero and its derivative is equal to zero and hence f is differentiable at the point x_0. □

Theorem 9.18. *Let f be a convex function and $x \in \operatorname{int}(\operatorname{dom} f)$. Then for the differentiability of the function f at a point x it is necessary and sufficient that its directional derivative $f'(x;\cdot)$ should be the linear function.*

Proof. The necessity is evident. Let us prove the sufficiency. For $y \in \mathbb{R}^n$, set $\phi(y) = f'(x;y)$. Under conditions of the theorem the function ϕ is linear and hence there exists $a \in \mathbb{R}^n$, for which we have $\phi(y) = \langle a, y\rangle\ \forall y$. The function ϕ is continuous. Hence, by Lemma 9.8 it is the support function for the set $A = \partial f(x)$. And by virtue of Corollary 8.16: $A = \{x\colon \langle x^*, x\rangle \leq \phi(x^*)\ \forall x^*\}$ yielding $A = \{a\}$. Therefore, the subdifferential $\partial f(x)$ contains a unique element and hence, by Theorem 9.17, the function f is differentiable at the point x. □

Let a proper convex function be given on $\mathbb{R}^n$. From the above mentioned, it is differentiable almost everywhere on $\operatorname{int}(\operatorname{dom} f)$. Moreover, by virtue of the remarkable theorem due to A. D. Aleksandrov,[4] each proper convex function given on $\mathbb{R}^n$ is twice differentiable almost everywhere on $\operatorname{int}(\operatorname{dom} f)$. Let us present the exact formulation.

Let D be a map assigning to each $x \in \mathbb{R}^n$ the square $n \times n$ symmetric matrix $D(x)$ with elements $d_{i,j}(x)$. The map D is called *locally summable* provided each of the functions $d_{i,j}(\cdot)$, $i, j = 1, \ldots, n$ is summable on every compact subset of $\mathbb{R}^n$. The collection of all such locally summable maps D will be denoted by $\mathcal{D}$.

Theorem 9.19 (A. D. Aleksandrov). *Let f be a proper convex function defined on $\mathbb{R}^n$. Then there exists such a locally summable map $D \in \mathcal{D}$, that for almost all $x \in \operatorname{int}(\operatorname{dom} f)$, the function f is differentiable at x and*

$$|f(y) - f(x) - \langle f'(x), (y-x)\rangle - \frac{1}{2}\langle D(x)(y-x), (y-x)\rangle| = o(|y-x|^2), \quad y \to x$$

holds true.

It is natural, in those points $x \in \operatorname{int}(\operatorname{dom} f)$ in which this relation is fulfilled, to call the matrix $D(x)$ the second derivative of the function f at the point x.

The proof of this theorem is rather complicated. It can be found in [34].

In conclusion of this section, let us consider the question about the approximation of a convex function by smooth convex functions. Let f be a continuous function given on a compact subset K of $\mathbb{R}^n$. Then by virtue of the classic Weierstrass theorem on approximation (see, e.g., [25]) there exists a sequence of infinitely differentiable

4 Aleksandr Danilovich Aleksandrov (1912–1999), a Russian mathematician.

on $\mathbb{R}^n$ functions f_i, uniformly convergent to the function f on the set K. Moreover, these functions may be chosen so that all of them are polynomials of x in a certain neighborhood of K.

In this connection the following natural question arises. If f is a proper convex function and $K \subset \operatorname{int}(\operatorname{dom} f)$, can the approximating smooth functions f_i be taken convex? The affirmative answer on this question gives the following assertion (it is Proposition 2 from [35]).

Theorem 9.20. *Let G be a convex bounded open set from $\mathbb{R}^n$ and f a convex proper function such that* $\operatorname{cl} G \subset \operatorname{int}(\operatorname{dom} f)$. *Then there exists such a sequence of convex functions f_i, that each function f_i is infinitely differentiable on the set G and the sequence $\{f_i\}$ uniformly converges to f on G.*

If we suppose, in addition, that the function f is k times continuously differentiable on G, then the sequence $\{f_i\}$ can be chosen so that all sequences of derivatives of the functions f_i converge to the corresponding derivatives of the function f uniformly on G.

Proof. Let us use some elements from the techniques employed in the theory of distributions (see [36], part II).[5] The function f is continuous on the set G. Take the δ-sequence of functions $\{\delta_i(x)\}$. Each of the functions δ_i is defined and infinitely differentiable on $\mathbb{R}^n$, nonnegative, vanishes for $|x| \geq 1/i$ and

$$\int_{\mathbb{R}^n} \delta_i(x)\, dx = 1 \quad \forall i.$$

Consider on the set G the functions $f_i(x) = f(x) * \delta_i(x)$, where the symbol $*$ denotes the convolution of functions, which is defined by the formula

$$f_i(x) = f(x) * \delta_i(x) = \int_{\mathbb{R}^n} f(x-\xi)\delta_i(\xi)\, d\xi.$$

By the properties of the functions δ_i, forming the δ-sequence and by the convexity of the function f, for arbitrary $x_1, x_2 \in G$ and $\alpha, \beta \geq 0 : \alpha + \beta = 1$, for each number i we have

$$\begin{aligned} f_i(\alpha x_1 + \beta x_2) &= \int_{\mathbb{R}^n} f(\alpha x_1 + \beta x_2 - \xi)\delta_i(\xi)\, d\xi \\ &\leq \int_{\mathbb{R}^n} (\alpha f(x_1 - \xi)\delta_i(\xi) + \beta f(x_2 - \xi)\delta_i(\xi))\, d\xi \\ &= \alpha f(x_1) * \delta_i(x_1) + \beta f(x_2) * \delta_i(x_2) = \alpha f_i(x_1) + \beta f_i(x_2). \end{aligned}$$

From here it follows that extending the function f_i outside the set G as $+\infty$, we get that each function f_i is convex. The uniform convergence of functions f_i to the function f on G as well as the convergence of corresponding derivatives are proved in [36]. □

5 The reader not familiar with this theory may omit this proof.

10 Convex cones

Let X be a real linear space.

Definition 10.1. A set $K \subseteq X$ is called the cone if for each $x \in K$ we have

$$\lambda x \in K \quad \forall \lambda > 0.$$

We distinguish an important class of *convex cones*.

Definition 10.2. A cone K is convex if for each $x_1, x_2 \in K$ we have

$$x_1 + x_2 \in K.$$

Consider some simplest examples of cones. Convex cones are: any linear subspace; nonnegative orthant in $\mathbb{R}^n$, which consists of all such vectors $x = (x_1, \ldots, x_n)$, that $x_i \geq 0\ \forall i$; the set of all continuous nonnegative functions from the space $C[a, b]$; the set of all nonnegatively defined matrices etc.

In the following, we will assume that X is the Euclidean space.

Definition 10.3. Let K be a cone. The cone

$$K^* = \{x^* \in X\colon \langle x^*, x\rangle \geq 0\ \forall x \in K\}$$

is called the conjugate cone to K.

Obviously, the conjugate cone K^* is always a cone, independently of the fact whether the cone K is convex or not. Zero belongs to the closure of each nonempty cone. Lastly, if $K_1 \subseteq K_2$, then $K_2^* \subseteq K_1^*$.

Notice that, parallel to the conjugate, *the normal cone* to the cone K which is denoted by K^0 and defined by the relation

$$K^0 = \{x^* \in X\colon \langle x^*, x\rangle \leq 0,\ \ \forall x \in K\}$$

is often considered. It is evident that $K^0 = -K^*$. The cone K^0 is also called *the polar cone*. We will restrict ourselves to considering only conjugate cones.

We will use the notation $K^{**} = (K^*)^*$. The cone K^{**} is called *the second conjugate to the cone K*.

Proposition 10.4. $K^{**} \supseteq K$.

Proof. Let $x \in K$. Then $\langle x, x^*\rangle \geq 0\ \forall x^* \in K^*$. Hence $x \in K^{**}$. □

Lemma 10.5. *Let K be a convex closed cone. Then $K^{**} = K$ or, in other words, it coincides with its second conjugate cone.*

DOI 10.1515/9783110460308-011

Proof. By virtue of Proposition 10.4 it is sufficient to prove that $K^{**} \subseteq K$. Let us do it. Take an arbitrary $y \notin K$. We prove the existence of such $x^* \in X$ that

$$\langle y, x^* \rangle < 0, \quad \langle z, x^* \rangle \geq 0 \; \forall z \in K. \tag{10.1}$$

Indeed, applying the theorem on strict separability to the convex closed set K and the point $\{y\}$, we obtain the existence of such $x^* \in X$, that

$$\langle y, x^* \rangle < \inf_{z \in K} \langle z, x^* \rangle.$$

Let us show that $\alpha = \inf_{z \in K} \langle z, x^* \rangle = 0$. In fact, $\alpha \leq 0$, since $0 \in K$. Suppose that $\alpha < 0$. Then there exists $z \in K$, for which $\langle z, x^* \rangle < 0$. So, $\alpha = -\infty$ since $yz \in K \; \forall y > 0$, contradicting the fact that, by construction, $\alpha > \langle y, x^* \rangle$. Therefore $\inf_{z \in K} \langle z, x^* \rangle = 0$ and hence relations (10.1) are proved.

From (10.1) we have $x^* \in K^*$. Hence $y \notin K^{**}$. By the arbitrariness of y we get the desired inclusion $K^{**} \subset K$. □

The cone represented in the form

$$K = \left\{ x \colon x = \sum_{i=1}^{m} \lambda_i a_i, \; \lambda_i \geq 0 \right\}, \tag{10.2}$$

where m is a positive integer and $a_i \in X$, $i = \overline{1, m}$ are given vectors, is called *finitely generated.*

Lemma 10.6 (On a finitely generated cone). *A finitely generated cone is closed.*

Proof. Apply the induction on m. Let $m = 1$. Then $K = \{x \colon x = \lambda a, \; \lambda \geq 0\}$. For the proof of the closedness of K it is sufficient to consider the case $a \neq 0$. Suppose that there exists a sequence $\{\lambda_i\}$ such that $\lambda_i a \to y$. Let us prove that $y \in K$.

For sufficiently large i we have $|\lambda_i a - y| \leq 1$, implying

$$|\lambda_i| \leq \frac{1 + |y|}{|a|}.$$

It means that the sequence $\{\lambda_i\}$ is bounded. Extracting from it the convergent subsequence, we will suppose that $\lambda_i \to \lambda_0 \geq 0$. Hence $\lambda_i a \to \lambda_0 a$. But $\lambda_i a \to y$ yielding $y = \lambda_0 a \in K$, completing the proof of the closedness of K for $m = 1$.

Suppose now that the desired assertion is true for $m - 1$. Let us prove it for m. Consider two cases. If $(-a_i) \in K$ for all numbers i, then K is the linear finite-dimensional subspace, whose closedness is well known from the course of functional analysis.

Consider the second case: there exists the number j, for which $(-a_j) \notin K$. Let, for definiteness, $(-a_m) \notin K$. Suppose that the sequence $\{x_i\}$ lies in K and $x_i \to x_0$. Prove that $x_0 \in K$.

Consider the cone $K_{m-1} = \{x \colon x = \sum_{i=1}^{m-1} \lambda_i a_i, \; \lambda_i \geq 0\}$. It is closed, by the assumption of induction. We have $K = K_{m-1} + \{x \colon x = \lambda_m a_m, \; \lambda_m \geq 0.\}$ Therefore, since $x_i \in K$ for all i, there exist $\tilde{x}_i \in K_{m-1}$ and $\lambda_{m,i} \geq 0$ such that $x_i = \tilde{x}_i + \lambda_{m,i} a_m$.

Let us show that the sequence $\{ \lambda_{m,i} \}$ is bounded. Suppose the contrary. Then, passing to a subsequence, we will suppose that $\lambda_{m,i} \to +\infty$. The sequence $\{x_i\}$ converges and hence it is bounded. Moreover, for large i we have $-a_m = \tilde{x}_i/\lambda_{m,i} - x_i/\lambda_{m,i}$, implying $\tilde{x}_i/\lambda_{m,i} \to (-a_m)$ and $\tilde{x}_i/\lambda_{m,i} \in K_{m-1}$. Therefore, by the closedness of the cone K_{m-1} we get $-a_m \in K_{m-1} \subset K$. This contradicts $(-a_m) \notin K$, yielding the boundedness of the sequence $\{ \lambda_{m,i} \}$.

So, the sequence $\{ \lambda_{m,i} \}$ is bounded. Extracting from it a convergent subsequence, we will assume, without loss of generality, that $\lambda_{m,i} \to \lambda_m \geq 0$. Hence $\tilde{x}_i = x_i - \lambda_{m,i} a_m \to x_0 - \lambda_m a_m$ implying that the sequence $\{ \tilde{x}_i \}$ converges. But since $\tilde{x}_i \in K_{m-1}$ for all i, and the cone K_{m-1} is closed by the induction assumption, then $\tilde{x}_i \to \tilde{x} \in K_{m-1}$ and hence $x_0 = \tilde{x} + \lambda_m a_m \in K$. The closedness of the cone K is proved, concluding the m-th step of the induction. □

Parallel with finitely generated cones, in applications so called *polyhedral cones* often arise. A cone is called *polyhedral* if it can be represented in the form

$$K = \left\{ x \in X \colon \langle a_i, x \rangle = 0, i = \overline{1, m_1}, \ \langle b_i, x \rangle \leq 0, i = \overline{1, m_2} \right\},$$

where m_1, m_2 are given positive integers and a_i, b_i are given vectors. It turns out that in finite-dimensional spaces there is a close relation between finitely generated and polyhedral cones. Namely, the following general assertion which is due to Weyl[1] and Minkowski holds true.

Theorem 10.7. *In a finite-dimensional space, each finitely generated cone is polyhedral and vise versa.*

The proof of this theorem may be found, e.g., in [2]. Notice that this theorem implies the validity of the lemma on a finitely generated cone since each polyhedral cone is obviously closed.

The calculation of conjugate cones is a very important problem in applications. We will present a few formulas which help to find conjugate cones in some cases. We will start with the assertion giving the formula for the conjugate cone in a particular, but important case. Let $X = \mathbb{R}^n$. For a vector $x \in \mathbb{R}^n$, we will write $x \geq 0$, if all its coordinates are nonnegative and $x \leq 0$, if they are nonpositive.

Lemma 10.8 (Farkas[2]). *Let*

$$K = \{ x \in \mathbb{R}^n : Ax = 0, \ Bx \geq 0 \},$$

where A, B are given matrices of dimensions $m \times n$ and $k \times n$ respectively. Then

$$K^* = D \colon = \{ y \in \mathbb{R}^n : y = A^T \lambda + B^T \mu, \ \lambda \in \mathbb{R}^m, \ \mu \in \mathbb{R}^k, \ \mu \geq 0 \},$$

where the symbol ^{T} denotes the transposition of matrix.

1 Hermann Klaus Hugo Weyl (1885–1955), a German mathematician and physicist.

2 Gyula Farkas (1847–1930), a Hungarian mathematician.

Proof. First, let us prove that $D \subseteq K^*$. Indeed, let $y \in D$. Then $y = A^T\lambda + B^T\mu$ for some $\lambda \in \mathbb{R}^m$, $\mu \in \mathbb{R}^k$, $\mu \geq 0$. For an arbitrary $x \in K$ we have

$$\langle y, x\rangle = \langle \lambda, Ax\rangle + \langle \mu, Bx\rangle = \langle \mu, Bx\rangle \geq 0,$$

since $Ax = 0$, $Bx \geq 0$, $\mu \geq 0$, and hence $y \in K^*$.

Let us prove the inverse inclusion: $K^* \subseteq D$. To do so, we will show first that $D^* \subseteq K$. In fact, let $x \in D^*$. Then $\langle x, A^T\lambda + B^T\mu\rangle \geq 0\ \forall\lambda,\ \forall\mu \geq 0$. From here, for $\mu = 0$ we get $\langle Ax, \lambda\rangle \geq 0\ \forall\lambda$, yielding $Ax = 0$. Similarly, for $\lambda = 0$ we have $\langle Bx, \mu\rangle \geq 0\ \forall\mu \geq 0$, implying $Bx \geq 0$. So, $x \in K$, proving the inclusion $D^* \subseteq K$, from which it follows that $K^* \subseteq D^{**}$. But, obviously, D is the convex cone, and moreover, it is easy to see that it is finitely generated. Therefore, by the lemma on a finitely generated cone, it is closed. Therefore, by Lemma 10.5 we get $D^{**} = D$, implying the desired inclusion $K^* \subseteq D$. □

Let us present two more useful formulas for the calculation of conjugate cones in a Hilbert space.

Lemma 10.9. *Let convex cones $K_1, K_2, \dots, K_m$ be given and the intersection of their interiors $\bigcap_{i=1}^m \operatorname{int} K_i$ is nonempty. Then*

$$\left(\bigcap_{i=1}^m K_i\right)^* = \sum_{i=1}^m K_i^*.$$

Proof. First, let us obtain a useful formula for indicator functions. Namely, if K is a convex cone, then for its indicator function δ_K the following formula holds true:

$$\delta_K^* = \delta_{(-K^*)}. \tag{10.3}$$

Indeed, for an arbitrary x^* we have

$$\delta_K^*(x^*) = \sup_{x\in X}(\langle x^*, x\rangle - \delta_K(x)) = \sup_{x\in K}\langle x^*, x\rangle,$$

yielding

$$\delta_K^*(x^*) = \begin{cases} 0, & x^* \in (-K^*) \\ +\infty, & x^* \notin (-K^*), \end{cases}$$

that proves formula (10.3).

Define convex functions f_i as $f_i = \delta_{K_i}$. By condition, $\bigcap_{i=1}^m \operatorname{int} K_i \neq \emptyset$. Therefore, all functions f_i vanish in a certain neighborhood of a point $x_0 \in \bigcap_{i=1}^m \operatorname{int} K_i$, and hence they are continuous in this neighborhood. So, for functions f_i all conditions of Theorem 7.13 are fulfilled, that implies

$$\delta_K^* = \delta_{K_1}^* \oplus \cdots \oplus \delta_{K_m}^*.$$

Here $K = \bigcap_{i=1}^m K_i$ and the evident equality $\delta_{K_1} + \cdots + \delta_{K_m} = \delta_K$ is used. From here, by virtue of (10.3) we have

$$\delta_{(-K^*)} = \delta_{(-K_1^*)} \oplus \cdots \oplus \delta_{(-K_m^*)}.$$

From the definitions of the infimal convolution and the indicator function it immediately follows that the infimal convolution of indicator functions of certain sets is equal to the indicator function of the sum of these sets and hence

$$\delta_{(-K_1^*)} \oplus \cdots \oplus \delta_{(-K_m^*)} = \delta_M,$$

where $M = (-K_1^*) + \cdots + (-K_m^*)$. Therefore, we proved that $\delta_{(-K^*)} = \delta_M$. But if the indicator functions of two sets are equal, then these sets coincide and hence $M = -K^*$, implying

$$(-M) = K_1^* + \cdots + K_m^* = K^* = \left(\bigcap_{i=1}^{m} K_i\right)^*,$$

that completes the proof. □

Notice, that another proof of this lemma, in some sense more natural and based on the separability theorem for convex cones is given in [12].

Exercise 10.10. *Let convex closed cones $K_1, K_2, \ldots, K_m$ be given.*

Prove that

$$\left(\sum_{i=1}^{m} K_i\right)^* = \bigcap_{i=1}^{m} K_i^*.$$

Decision. Let $y \in \cap_{i=1}^m K_i^*$. Then $\langle y, x_i \rangle \geq 0$ for all $x_i \in K_i$ and $i = 1, \ldots, m$. From here we obtain that $\langle y, x \rangle \geq 0$ for all x such that $x = x_1 + \cdots + x_m$, $x_i \in K_i$ $\forall i$ and hence $y \in (K_1 + \cdots + K_m)^*$.

The inverse inclusion follows from the fact that $K_i \subseteq (\tilde{K}_1 + \cdots + \tilde{K}_m)$ for all i and hence $(\tilde{K}_1 + \cdots + \tilde{K}_m)^* \subseteq K_i^*$ for all i, where $\tilde{K}_i = K_i \bigcup \{0\}$. Therefore $(K_1 + \cdots + K_m)^* \subseteq \bigcap_{i=1}^m K_i^*$, since, as it is easy to see $(K_1 + \cdots + K_m)^* = (\tilde{K}_1 + \cdots + \tilde{K}_m)^*$.

11 A little more about convex cones in infinite-dimensional spaces

Throughout this section we will assume that X is a Banach space and K is a convex closed cone in it, such that $K \neq X$. In this situation we will be interested in the case when the space X is infinite-dimensional since otherwise the results given below immediately follow from the finite-dimensional separability of convex sets.[1]

The following question is of great interest for applications: whether there exists a linear continuous functional $l \neq 0$ on X which is nonnegative on the cone K, i.e., $\langle l, x \rangle \geq 0$ for all $x \in K$, or even positive on K:

$$\langle l, x \rangle > 0 \quad \forall x \in K \setminus \{0\}. \tag{11.1}$$

The existence of such linear functionals means the possibility to separate the cone K from zero or strictly separate the cone $K \setminus \{0\}$ from zero in the sense of (11.1). However, as mentioned above, for the separability of convex sets in infinite-dimensional spaces it is important that either the interior of at least one of these sets is nonempty or one of them is closed and the other is compact. It turns out that it is possible to avoid, in some sense, these restrictive conditions for cones. Let us present the corresponding assertions.

Theorem 11.1. *For an arbitrary $x_0 \notin K$ there exists such a linear continuous functional l which is nonnegative on K and, moreover, $\langle l, x_0 \rangle < 0$.*

Proof. Let $x_0 \notin K$. Then the convex compact set $\{x_0\}$ and the convex closed set K may be strictly separated. Hence there exists such a linear continuous functional l, that $\langle l, x \rangle > \langle l, x_0 \rangle$ for all $x \in K$. It is obvious that $\langle l, x_0 \rangle < 0$, since $0 \in K$ and $\langle l, 0 \rangle = 0$. Take an arbitrary $x \in K$ and show that $\langle l, x \rangle \geq 0$. Indeed, suppose that $\langle l, x \rangle < 0$. Then obviously $\lambda x \in K$ for all $\lambda > 0$ and $\langle l, (\lambda x) \rangle \to -\infty$ for $\lambda \to \infty$, giving the contradiction. □

Definition 11.2. A cone K is acute if it does not contain one-dimensional subspaces or, in other words, if $x \in K$ and $x \neq 0$, then $(-x) \notin K$.

Theorem 11.3. *Let K be a convex, closed, and acute cone. Then for each vector $x \in K \setminus \{0\}$ there exists a linear continuous functional l which is nonnegative on the cone K and $\langle l, x \rangle > 0$.*

Proof. Let $x \in K \setminus \{0\}$. Then $(-x) \notin K$, since the cone K is acute. Applying the previous theorem to the point $x_0 = -x$ we obtain the desired result. □

---

1 A reader who does not take an interest in infinite-dimensional specificity may omit this section without prejudice to further understanding.

DOI 10.1515/9783110460308-012

Remember that a normed space X is called *separable* provided it contains a countable, everywhere dense, subset Ξ, i.e., for each point $x \in X$ there exists a sequence convergent to x and all its elements belong to the set Ξ.

Theorem 11.4. *Let a Banach space X be separable and K a convex closed and acute cone in it. Then there exists a linear continuous functional which is positive on the cone K, i.e., $\langle l, x\rangle > 0$ for all $x \in K \setminus \{0\}$.*

The proof of this theorem can be found in [37], § 2. It is important to note that the assumption about the separability of the space X is essential. Indeed, let T be an arbitrary metric space and $X = C(T)$ a linear space of continuous bounded functions $f : T \to \mathbb{R}$ with the norm of the uniform convergence $\|f\| = \sup\{|f(t)| : t \in T\}$. As is well known, the space X is Banach with respect to this norm.

Denote by K the set of all nonnegative functions $f \in C(T)$. It is clear that K is a convex closed cone. In [37], § 2 it is proved that if a metric space T is not separable, then there does not exist a linear (even discontinuous) functional that is positive on a cone K. The crucial fact is that in this case the space $C(T)$ is not separable.

In conclusion, let us discuss which of the known Banach spaces are separable and which are not. Among the separable Banach spaces are: $\mathbb{R}^n$; $C[a, b]$, the space of continuous functions on an interval $[a, b]$ with the norm of uniform convergence; for $1 \le p < +\infty$, the space $L_p[a, b]$ of functions f measurable on $[a, b]$, for which the function $|f|^p$ is summable, with the norm $\|f\| = (\int_a^b |f(x)|^p\,dx)^{1/p}$; and also the space l_p, consisting of real sequences $x = (x^1, x^2, \ldots)$, for which the series $\sum_{i=1}^{\infty} |x^i|^p$ converges, with the norm $\|x\| = \left(\sum_{i=1}^{\infty} |x^i|^p\right)^{1/p}$ (see, e.g., [27]).

Let us present examples of nonseparable Banach spaces. We may consider: the space l_∞ of bounded sequences $x = (x^1, x^2, \ldots)$ with the norm $\|x\| = \sup_i \{|x^i|\}$, and the space $L_\infty[a, b]$ of measurable essentially bounded functions f with the norm $\|f\| = \operatorname{ess\,sup} |f(x)|$.

The proof of nonseparability of these spaces is based on the following evident assertion. Consider a family of elements x_α of the space X, depending on the parameter α, taking its values in a given set A of a continuum power (for example, $A = [0, \alpha_0]$ for some $\alpha_0 > 0$). Suppose that there exists such an $\varepsilon > 0$, that $\|x_\alpha - x_\beta\| \ge \varepsilon$ for each $\alpha, \beta \in A$ with $\alpha \ne \beta$. Then the space X is nonseparable. In the space l_∞ the corresponding family is formed by the set of all possible sequences x, consisting of zeroes and units (as is known, this set is of continuum power), and in the space $L_\infty[a, b]$ it is the family of characteristic functions χ_t of the interval $[a, t]$, where $t \in [a, b]$.

Consider the Banach space $C(\mathbb{R})$ of continuous bounded functions on $\mathbb{R}$ with the norm $\|f\| = \sup |f(x)|$. Let us prove that this space is also nonseparable. Indeed, define the function $\varphi \in C(\mathbb{R})$ by the relations

$$\varphi(x) = \begin{cases} nx + 1 - n^2, & x \in [n - 1/n, n], \\ -nx + 1 + n^2, & x \in [n, n + 1/n], \\ 0, & x \notin \bigcup_{n=1}^{\infty} [n - 1/n, n + 1/n], \end{cases}$$

where n takes natural values $n = 1, 2, \ldots$. Obviously, $\|\varphi\| = 1$. For $\alpha \in A = [0, 1/2]$, set $\varphi_\alpha(x) = \varphi(x + \alpha)$, $x \in \mathbb{R}$. It is clear that $\varphi_\alpha(n - \alpha) = 1$ for each n. In addition, for each $\beta \in [0, 1/2]$, $\beta \neq \alpha$ we have $\varphi_\beta(n - \alpha) = \varphi(n + (\beta - \alpha)) = 0$, if an integer n is chosen so that $|\alpha - \beta| > 1/n$. Hence, $\|\varphi_\alpha - \varphi_\beta\| = 1$ for each $\alpha, \beta \in A\colon\ \alpha \neq \beta$ and, hence, the family of functions φ_α, $\alpha \in A$ gives the desired continual family of elements of the space $C(\mathbb{R})$, for which the distance between every two distinct elements equals one that shows its nonseparability.

12 A problem of linear programming

The study of extremal problems with constraints (i.e., problems of optimization of a given function on a certain subset of the initial space) plays a great role in contemporary mathematics. The problem of linear programming occupies a very special place. It consists in the optimization of a linear function on a set of points satisfying a finite number of linear constraints taking the form of equalities or inequalities. The linear programming problem differs from other extremal problems in, on the one hand, that it often arises in applications and, on the other hand, that it is the most extensively studied problem and effective numerical methods for its solution have been developed. Here we only get acquainted with the elements of this theory. For a more detailed study we refer the reader to the corresponding extensive literature, from which we can select, e.g., [38].

Let us present one of possible formulations of the linear programming problem. Let vectors $c \in \mathbb{R}^n$, $b \in \mathbb{R}^m$ and a linear operator (matrix) $A: \mathbb{R}^n \to \mathbb{R}^m$ be given. Consider *the problem of linear programming*:

$$\langle c, x\rangle \to \min, \quad x \in M = \{x\colon Ax \geq b,\ x \geq 0\}, \tag{12.1}$$

which consists in the minimization of a linear function $\langle c, x\rangle$ on the set of all admissible points M.

It is easy to prove that the set of all admissible points M is convex and closed, but it may happen to be empty or unbounded. So, respectively, the same may also be said about the solutions set of the linear programming problem.

We will start with the existence theorem of linear programming. It turns out that if the set of admissible points is nonempty and the minimizing function is bounded below on it, then the minimum in this problem can be achieved. Namely, the following result holds true.

Theorem 12.1 (On the existence of a solution). *Let $M \neq \varnothing$ and*

$$y\colon = \inf_{x \in M} \langle c, x\rangle > -\infty.$$

Then there exists $x_0 \in M$ such that $\langle c, x_0\rangle = y$.

Proof. Consider the set K in the space $\mathbb{R}^{m+1}$ consisting of vectors (α, z), $\alpha \in \mathbb{R}$, $z \in \mathbb{R}^m$, for which there exists a vector $x \in \mathbb{R}^n$, satisfying the inequalities

$$\alpha \geq \langle c, x\rangle \quad x \geq 0, \quad Ax \geq z.$$

It is easy to see that K is the cone and, moreover, it is generated by the vectors

$$\bar{a}_1, \ldots, \bar{a}_n, e_1, -e_2, \ldots, -e_{m+1}.$$

Here $e_i \in \mathbb{R}^{m+1}$, $i = \overline{1, m+1}$ denotes the vector which has the unity on the i-th position and zeros on all others, and $\bar{a}_i = (c_i, a_i) \in \mathbb{R}^{m+1}$, $i = \overline{1, n}$, where c_i is the i-th coordinate of the vector c, and a_i is the i-th column of the matrix A.

DOI 10.1515/9783110460308-013

By virtue of the lemma on finitely generated cones, K is the closed cone. Consider the set

$$C = \{ \alpha \in \mathbb{R}\colon (\alpha, b) \in K \}.$$

If $x \in M$ and $\alpha = \langle c, x \rangle$, then $(\alpha, b) \in K$ and, hence, $\alpha = \langle c, x \rangle \in C$. So, in particular, the set C is nonempty. Moreover, C is obviously closed and $\alpha \geq \gamma \ \forall \alpha \in C$. The latter follows from the fact that if $\alpha \in C$, then $(\alpha, b) \in K$ and hence there exists such an admissible point $x \in M$ that $\alpha \geq \langle c, x \rangle$.

So, from the above mentioned it follows that $\gamma = \inf \{ \alpha \in C \}$, and, moreover, this infimum is achieved. Therefore, $\gamma \in C$, implying $(\gamma, b) \in K$. Then there exists such an admissible point $x \in M$, that $\gamma = \langle c, x \rangle$ and, hence, x is the solution of the initial linear programming problem. □

One of the most effective methods of solving the linear programming problem consists in investigating, parallel to the initial problem (12.1), the so-called *dual problem* which is also a problem of linear programming. It is useful both from the theoretical point of view as well as for the realization of numerical methods for its solution (see [39]). Let us turn to the construction of the dual problem.

The cone K, constructed above during the proof of the existence problem, is convex, closed and satisfies the following property: if $(\alpha, z) \in K$ and $\beta \geq \alpha$, then $(\beta, z) \in K$. Therefore, K is the epigraph of the convex closed function

$$S(z) = \inf \{ \alpha \in \mathbb{R}\colon (\alpha, z) \in K \}.$$

From the proof of the existence problem it follows that $S(z)$ is the minimum in the linear programming problem

$$\langle c, x \rangle \to \min, \quad Ax \geq z, \ x \geq 0, \tag{12.2}$$

which is often called the perturbation of initial problem (12.1), since in this problem the vector b is replaced with the vector z, playing the role of a parameter.

Evaluate the function, conjugate to S. We have

$$\begin{aligned} S^*(y) &= \sup_z \{ \langle y, z \rangle - S(z) \} \\ &= \sup_z \{ \langle y, z \rangle - \inf \{ \langle c, x \rangle : x \in \mathbb{R}^n, \ x \geq 0, \ Ax \geq z \} \} \\ &= \sup \{ \langle y, z \rangle - \langle c, x \rangle : x \in \mathbb{R}^n, \ z \in R^m, \ x \geq 0, \ Ax \geq z \}. \end{aligned}$$

Obviously,

$$\sup \{ \langle y, z \rangle : z \leq Ax \} = \langle Ax, y \rangle < +\infty$$

if and only if $y \geq 0$. Therefore,

$$S^*(y) = \begin{cases} \sup_{x \geq 0} \langle A^* y - c, x \rangle, & \text{if } y \geq 0 \\ +\infty, & \text{otherwise} \end{cases}$$

and hence

$$S^*(y) = \begin{cases} 0, & \text{if } A^*y \le c,\ y \ge 0 \\ +\infty, & \text{otherwise.} \end{cases}$$

Therefore

$$S^{**}(z) = \sup\{\langle y, z\rangle : A^*y \le c,\ y \ge 0\}.$$

Particularly, $S^{**}(b)$ coincides with the value of the maximum in the problem

$$\langle b, y\rangle \to \max, \quad y \in \mathbb{R}^m,\ y \ge 0,\ A^*y \le c, \tag{12.3}$$

consisting in the maximization of the linear function $\langle b, y\rangle$ on the set of admissible points $N = \{y \in \mathbb{R}^m : y \ge 0,\ A^*y \le c\}$. This problem is called *dual to the initial linear programming problem* (12.1).

Notice that problem (12.3) is also the linear programming problem, and moreover, the initial problem (12.1) is dual to it.

That is why problems (12.1) and (12.3) are called the pair of mutually dual problems. The close connection between these problems is illustrated by the following assertion. To formulate this result, let us mention that the infimum of the function $\langle c, x\rangle$ on the set M will be called *the value of problem* (12.1). Respectively, the supremum of the function $\langle b, y\rangle$ on the set N is called *the value of dual problem* (12.3).

Theorem 12.2 (The duality theorem). *For a pair of mutually dual problems the following alternative holds true: either the values of both problems are finite and equal and in both problems there exists a solution or in one of these problems the set of admissible points is empty whereas in the dual problem the set of admissible points is empty or the value of this problem is infinite.*

Proof. We will give the proof following [7]. First let the function S is proper. By that mentioned above, the function S is closed. Therefore, by the Fenchel–Moreau theorem we have $S(b) = S^{**}(b)$. If also $b \in \operatorname{dom} S$, then the values of both problems are finite and equal and, by the previous theorem, the solutions of these problems exist. From the other side, if $b \notin \operatorname{dom} S$, then $S(b) = S^{**}(b) = +\infty$. Hence, in particular, in the initial problem (12.1) the set of admissible points is empty.

Suppose now that S is not a proper function. If, additionally, $S(z) = \pm\infty\ \forall z$, then $S^{**}(z) = \pm\infty\ \forall z$ and we get the above considered case. Suppose the contrary. Then $\operatorname{dom} S \ne \emptyset$. Hence $S^{**} \equiv -\infty$ and, therefore $S^{**}(b) = -\infty$. It means that the set of admissible points in the dual problem (12.3) is empty. Then in initial problem (12.1), this set is either empty (when $S(b) = +\infty$), or a minimized function is unbounded below on the set of admissible points (when $S(b) = -\infty$). □

Let us present necessary and sufficient conditions for the optimality of the considered linear programming problem.

Theorem 12.3. *A point $x^* \in \mathbb{R}^n$, which is admissible in initial problem* (12.1) *is its solution if and only if there exists such a vector $y^* \in \mathbb{R}^m$, that the following holds*

$$c \geq A^* y^*, \quad y^* \geq 0, \ \langle c - A^* y^*, x^* \rangle = 0, \ \langle Ax^* - b, y^* \rangle = 0. \tag{12.4}$$

Moreover, from (12.4) *it follows that y^* is a solution of the dual problem* (12.3).

An analogous assertion is true for the dual problem (12.3) *also.*

Proof. Necessity. Let x^* be a solution of problem (12.1).

Consider the case

$$K = \{x \colon x_i \geq 0 \ \forall i \colon x_i^* = 0, \quad (Ax)_i \geq 0 \ \forall i \colon (Ax^* - b)_i = 0\},$$

where, as we can recall, the lower index i denotes the i-th coordinate of a vector. It is easy to see that $\langle c, x \rangle \geq 0 \ \forall x \in K$, implying $c \in K^*$. Hence, by the Farkas lemma, there exist $y^* \in \mathbb{R}^m$, $\mu^* \in \mathbb{R}^n$ such that

$$c = A^* y^* + \mu^*, \quad y^* \geq 0, \ \mu^* \geq 0, \ \langle Ax^* - b, y^* \rangle = 0, \ \langle \mu^*, x^* \rangle = 0.$$

From here, since $\mu^* = c - A^* y^*$, we obtain (12.4). Further, by virtue of (12.4) the point y^* is admissible for the dual problem (12.3) and, moreover, $\langle b, y^* \rangle = \langle Ax^*, y^* \rangle = \langle c, x^* \rangle$, implying $\langle b, y^* \rangle = \langle c, x^* \rangle$ and hence, by virtue of the duality theorem, y^* is the solution of the duality problem (12.3).

Sufficiency. Let x^* be an admissible point of problem (12.1) and there exists y^*, for which (12.4) holds. Then, from that mentioned above it follows that $\langle b, y^* \rangle = \langle c, x^* \rangle$. Let x be an arbitrary admissible point in problem (12.1), i.e., $x \geq 0$, $Ax \geq b$. Then, by virtue of (12.4), we get

$$\begin{aligned} c \geq A^* y^* \Rightarrow \langle c, x \rangle &\geq \langle A^* y^*, x \rangle = \langle y^*, Ax \rangle \\ &\geq \langle y^*, b \rangle = \langle c, x^* \rangle \Rightarrow \langle c, x \rangle \geq \langle c, x^* \rangle \end{aligned}$$

and hence x^* is a solution of problem (12.1). □

13 More about convex sets and convex hulls

Let X be a normed space. Recall that a point x is called *a boundary point of a set $A \subset X$*, if there exist such sequences $\{x_i\}$ and $\{x_i'\}$ convergent to it that $x_i \in A,\ x_i' \notin A\ \forall i$. The set of all boundary points of the set A is called its *boundary* and it is denoted by ∂A. If A is closed then the following is true: $\operatorname{int} A = A \setminus \partial A$.

Introduce one of the most important notions of a convex analysis, the notion of *an extreme point of a convex set*. Let $A \subset X$ be a convex closed set.

Definition 13.1. A point $x_0 \in A$ is called an extreme point of the set A, if such points $x_1, x_2 \in A$, $x_1 \neq x_2$ do not exist that $x_0 \in (x_1, x_2)$ (i.e., $x_0 = \alpha x_1 + (1-\alpha)x_2$ for a certain $\alpha\colon 0 < \alpha < 1$). The set of all extreme points of the set A is denoted by $\operatorname{ext}(A)$.

If A consists of a single point, then $A = \operatorname{ext}(A)$. Each extreme point of a set is its boundary point, but the converse is, naturally, not true. Moreover, if a point x belongs to the relative interior of a set containing more than one set, then x is not its extreme point.

If A is a nonzero subspace or semi-space, then the set of its extreme points is empty. From the other side, if a set A is compact, then the latter is impossible. Namely, the following important theorem holds true.

Theorem 13.2 (Krein[1]–Milman[2]). *Let $A \subset X$ be a compact set. Then the set of its extreme points* $\operatorname{ext}(A)$ *is nonempty and, moreover, the following formula holds true:*

$$A = \operatorname{cl}(\operatorname{conv}(\operatorname{ext}(A))). \tag{13.1}$$

The proof of this theorem in the general case needs more refined techniques of functional analysis than those that we are using, and it can be found, for example, in [30]. We will prove this theorem in a finite-dimensional case. But first let us mention the following interesting fact.

Even if A is a convex compact set in a finite-dimensional space, the set of its extreme points $\operatorname{ext}(A)$, which is nonempty by the Krein–Milman theorem, may be nonclosed. Here is the corresponding example.

Let $X = \mathbb{R}^3$, $A_1 = \{x = (x_1, x_2, x_3)\colon x_1^2 + x_2^2 = 1,\ x_3 = 0\}$ be a unit circle, $A_2 = \{x\colon x_1 = 1, x_2 = 0, |x_3| \le 1\}$ a vertical interval and $A = \operatorname{conv}(A_1 \bigcup A_2)$. By Theorem 2.4 the set A is compact. It is easy to see that $A_1 \setminus \{(1,0,0)\} \subset \operatorname{ext}(A)$, however, $(1,0,0) \notin \operatorname{ext}(A)$. Therefore, the set $\operatorname{ext}(A)$ is not closed.

If the space X is infinite-dimensional, then the operation of the closure in equality (13.1) cannot be omitted. However in a finite-dimensional case it can be done.

1 Mark Grigorievich Krein (1907–1989), a Russian mathematician.

2 David Pinhusovich Milman (1912–1982), a Russian mathematician.

DOI 10.1515/9783110460308-014

Theorem 13.3. *Let $A \subset \mathbb{R}^n$ be a convex compact set. Then the set of its extreme points* $\operatorname{ext}(A)$ *is nonempty and, moreover, the following equality holds true:*

$$A = \operatorname{conv}(\operatorname{ext}(A)).$$

Proof. We will use the induction in the dimension $\dim A$. If $\dim A = 0$, the assertion is evident. Let it be true for sets of the dimension less then m and suppose $\dim A = m$. Without loss of generality (i.e., by shifting the set A to zero and passing from $\mathbb{R}^n$ to the linear hull of this set), we may assume that $m = n$.

Take an arbitrary boundary point $a \in \partial A$. By virtue of the theorem on a finite-dimensional separability for convex sets there exists such an $x^* \in \mathbb{R}^n$, $x^* \neq 0$, that

$$\langle x^*, a\rangle \geq \langle x^*, x\rangle \quad \forall x \in A.$$

Set

$$\gamma = \langle x^*, a\rangle, \quad B = \{x \in A : \gamma = \langle x^*, x\rangle\}.$$

It is evident that B is a compact convex set, $B \subset A$ and $\dim B < \dim A$, since, by construction, $\dim A = n$ and $\dim B \leq n - 1$. So, by the induction assumption, $B = \operatorname{conv}(\operatorname{ext}(B))$ and, hence, $a \in \operatorname{conv}(\operatorname{ext}(B))$.

Let us show that $\operatorname{ext}(B) \subset \operatorname{ext}(A)$. Indeed, let $\xi \in \operatorname{ext}(B) \setminus \operatorname{ext}(A)$. Then there exist $x_1, x_2 \in A$, $x_1 \neq x_2$ and $\alpha \in (0,1)$ such that

$$\xi = \alpha x_1 + (1-\alpha)x_2.$$

Then

$$\langle x^*, x_1\rangle \leq \gamma, \quad \langle x^*, x_2\rangle \leq \gamma,$$

implying

$$\gamma = \langle x^*, a\rangle = \alpha\langle x^*, x_1\rangle + (1-\alpha)\langle x^*, x_2\rangle \leq \gamma\alpha + (1-\alpha)\gamma = \gamma.$$

But the latter is possible only in the case when

$$\gamma = \langle x^*, x_1\rangle = \langle x^*, x_2\rangle$$

and hence $x_1, x_2 \in B$, contrary to the assumption that $\xi \in \operatorname{ext}(B)$.

So, $\operatorname{ext}(B) \subset \operatorname{ext}(A)$ and hence $\operatorname{conv}(\operatorname{ext}(B)) \subset \operatorname{conv}(\operatorname{ext}(A))$, implying $a \in \operatorname{conv}(\operatorname{ext}(A))$. Therefore it is proved that

$$\partial A \subset \operatorname{conv}(\operatorname{ext}(A)).$$

Now, let $x \in A \setminus \partial A$. Then $x \in \operatorname{int} A$. Consider any direct line passing through the point x. It is easy to see that this line intersects the border ∂A of the set A at two points $x_1, x_2 \in \partial A$. By virtue of that proved above $x_1, x_2 \in \operatorname{conv}(\operatorname{ext}(A))$ and hence $x \in \operatorname{conv}(\operatorname{ext}(A))$, since $x \in [x_1, x_2]$. □

Now, let us turn to the Carathéodory theorem proved above. Recall it. Let A be an arbitrary set from $\mathbb{R}^n$. Then its convex hull conv A consists of all possible convex combinations of no more than $n + 1$ points of the set A. We ask, can we diminish the number of these points to n? Evident examples show that it is impossible without additional assumptions imposed on the set A. But, for example, if the set A is the cone and $0 \in A$, then to form its convex hull, it is sufficient to take only all possible convex combinations of linearly independent vectors of the set A (and their number is evidently not greater than n). Let us prove this assertion in analogy with the proof of the Carathéodory theorem given above.

Indeed, let A be an arbitrary cone in $\mathbb{R}^n$ and $0 \in A$. Let $x \neq 0$, $x \in \operatorname{conv} A$. Then, by Theorem 2.2, for some positive integer r the following representation takes place: $x = \sum_{i=1}^r \alpha_i x_i$, where $\alpha_i > 0$, $\sum_{i=1}^r \alpha_i = 1$, $x_i \in A\ \forall i$. Suppose that the vectors $x_1, \ldots, x_r$ are linearly dependent. Show that in this case x may be represented in the form of a convex combination of no more than $(r-1)$ points of the set A.

By virtue of the linear dependence of vectors $x_1, \ldots, x_r$ there exist such numbers $t_1, \ldots, t_r$, not all equal to zero, that

$$\sum_{i=1}^r t_i x_i = 0.$$

Without loss of generality, we will assume that there is at least one negative number among the numbers t_i (otherwise we will replace all of them with $(-t_i)$).

For an arbitrary number c we have the following:

$$x = \sum_{i=1}^r \alpha_i x_i + \sum_{i=1}^r c\, t_i x_i = \sum_{i=1}^r (\alpha_i + c\, t_i) x_i. \tag{13.2}$$

Since $\alpha_i > 0\ \forall i$, we get that for $c = 0$ all numbers $(\alpha_i + c\, t_i)$ are positive. Let us increase the parameter c from zero to $+\infty$. Since at least one of the numbers t_i is negative, then evidently there exists a smallest number $c > 0$ such that $(\alpha_i + c\, t_i) \geq 0\ \forall i$ and for a certain number $i_0 \leq r$ we have $(\alpha_{i_0} + c\, t_{i_0}) = 0$. So, omitting i_0-th term $(\alpha_{i_0} + c\, t_{i_0}) x_{i_0}$ in the representation (13.2), we get the desired assertion, i.e., that x may be represented in the form of a convex linear combination of no more than $(r-1)$ points of the set A, namely

$$x = \sum_{i \neq i_0} \tilde{\alpha}_i \tilde{x}_i,$$

where

$$\tilde{\alpha}_i = (\alpha_i + c t_i)/\tilde{\alpha} \geq 0, \quad \tilde{x}_i = \tilde{\alpha} x_i, \quad \tilde{\alpha} = \sum_{i=1}^r (\alpha_i + c\, t_i) > 0.$$

Notice that $\sum_{i \neq i_0} \tilde{\alpha}_i = 1$ and also $\tilde{x}_i \in A\ \forall i$, since A is the cone.

Repeating this procedure a finite number of times, we will complete the proof of the desired assertion.

Let us present a more profound assertion containing a condition on the set A which also guarantees that for the construction of its convex hull it is sufficient to take all possible convex combinations of collections consisting of no more than n of its points. It is the generalized Carathéodory theorem (see [40]), the proof of which is based on the Carathéodory theorem itself. First let us recall the notion of an arcwise connected set.

Definition 13.4. A subset A of a normed space X is called arcwise connected if each of its two points a, b may be connected by a continuous curve lying in A. The latter means that there exists such a continuous function $\varphi\colon [0,1] \to A$, that $\varphi(0) = a$, $\varphi(1) = b$.

Theorem 13.5 (A generalized Carathéodory theorem). *Let $A \subset \mathbb{R}^n$ be the union of no more than n arcwise connected sets. Then* conv A *consists of all possible convex combinations of no more than n points of the set A.*

Proof. Take an arbitrary point $y \in \operatorname{conv} A$ and prove that it can be represented in the form of a convex combination of no more than n points of the set A. Without loss of generality, we will assume that $y = 0$. By the Carathéodory theorem there exist points $x_i \in A$ and $\alpha_i \geq 0$, $i = \overline{1, n+1}$ such that

$$\sum_{i=1}^{n+1} \alpha_i = 1, \quad 0 = \sum_{i=1}^{n+1} \alpha_i x_i.$$

If at least one of the numbers α_i is equal to zero then the desired assertion is proved. So, we will assume that $\alpha_i > 0$ for all i. We will assume also that the vectors $x_1, \ldots, x_{n+1}$ are affine independent, since otherwise all of them belong to a proper subspace and the desired assertion follows immediately from the fact that

$$0 \in \operatorname{conv}\{x_1, \ldots, x_{n+1}\},$$

and the Carathéodory theorem.

For each $i \in \{1, \ldots, n+1\}$ set $x_i' = -x_i$ and denote by K_i the cone spanned over the vectors $\{x_j'\}$, $j \neq i$. Then $x_i \in K_i$, since for each i we have

$$x_i = \alpha_i^{-1} \sum_{j \neq i} \alpha_j x_j'.$$

Moreover, from the affine independence of systems of vectors $x_1, \ldots, x_{n+1}$ it follows that $x_i \notin K_j$ for all $i \neq j$, the interior of each of cones K_i is nonempty and these interiors are disjoint.

The number of points x_i equals $n + 1$, and the set A is the union of no more than n arcwise connected sets. Therefore, at least in one of these arcwise connected sets, say in $A_1 \subset A$, we can find at least two of the mentioned points. For determinacy, let $x_1, x_2 \in A_1$. By virtue of the arcwise connectedness of the set A_1 these points may be joined by a continuous curve lying in this set. But by virtue of the above mentioned, we have $x_1 \in K_1$, $x_2 \notin K_1$, from which it easily follows that the curve intersects the boundary of the cone K_1 at a certain point p. Obviously, $p \in A_1 \subset A$ and, since the vector p also belongs to the boundary of the cone K_1, then it can be represented in the form

$$p = \sum_{i=2}^{n+1} \mu_i x_i', \quad \mu_i \geq 0 \ \forall i,$$

where at least one of the numbers μ_i vanishes. For determinacy, let $\mu_2 = 0$. Then, obviously, $0 \in \operatorname{conv}\{p, x_3, \dots, x_{n+1}\}$, i.e., zero is the convex combination of no more than n points of the set A. □

In conclusion, we present the assertion which connects the property of local convexity of a set with its global property of convexity.

Let X be a normed space.

Definition 13.6. A set $A \subset X$ is called locally convex if for each $x \in A$ there exists $\varepsilon > 0$ such that the set $O(x, \varepsilon) \bigcap A$ is convex.

Obviously each convex set is locally convex but the converse is surely not true.

Definition 13.7. A set $A \subset X$ is connected if such nonempty subsets A_1, A_2 and open sets O_1, O_2 do not exist that

$$A_1 \subset O_1, \quad A_2 \subset O_2, \quad O_1 \bigcap O_2 \bigcap A = \emptyset, \quad O_1 \bigcup O_2 \supset A.$$

Notice that each arcwise connected set is connected but the converse is not true.

Lemma 13.8. *Suppose that a set $A \subset \mathbb{R}^2$ is locally convex, compact and points a, b belong to A. Let there exist a continuous map $\xi\colon [0, 1] \to \mathbb{R}^2$ such that $\xi(0) = a$, $\xi(1) = b$, and $\xi(\alpha) \in A$ for all $\alpha \in [0, 1]$. Then*

$$\alpha a + (1 - \alpha) b \in A \quad \forall \alpha \in [0, 1].$$

Proof. A detailed proof of this assertion can be found in [35]. We will outline its main steps.

First it can be shown that there exists a polygonal line with a finite number of components which connects the points a and b and lies in A. Then, by the theorem on the existence of a shortest line (see, e.g., [27]), points a and b may be connected in A by a continuous curve ω of a minimal length with the radius-vector $\eta(t)$, $t \in [0, 1]$. Let this curve be not an interval. It can be proved that then there exist numbers $t_0, t_1, t_2 \in (0, 1)$ and $\varepsilon > 0$ such that $t_1 < t_0 < t_2$, the set $O(\eta(t_0), \varepsilon) \cap A$ is convex, points $\eta(t_1)$ and $\eta(t_2)$ belong to the neighborhood $O(\eta(t_0), \varepsilon)$, however, the points $\eta(t_1), \eta(t_0), \eta(t_2)$ do not lie on a direct line.

Define the curve $\tilde{\eta}(t)$, $t \in [0, 1]$: for $t \in [0, t_1] \cup [t_2, 1]$ we have $\tilde{\eta}(t) = \eta(t)$, and for $t \in [t_1, t_2]$ the radius-vector $\tilde{\eta}(t)$ circumscribes the interval connecting the points $\eta(t_1), \eta(t_2)$. This curve completely lies in A and has length shorter than that of the curve ω. The obtained contradiction proves the desired assertion. □

Theorem 13.9. *Let a set $A \subset X$ be connected, locally convex and closed. Then A is convex.*

Proof. Take a point $a \in A$. Set

$$C = \{x : x \in A,\ \alpha a + (1 - \alpha)x \in A\ \forall \alpha \in [0, 1]\}.$$

Let us show that the set C is closed. Let a point b belong to a set A but not to the set C. Take $\varepsilon > 0$ such that the set $O(b, \varepsilon) \bigcap A$ should be convex. Suppose that there exists a point $d \in O(b, \varepsilon) \bigcap C$. Then $\alpha b + (1 - \alpha)d \in A$ and $\alpha a + (1 - \alpha)d \in A$ for all $\alpha \in [0, 1]$. For the set $R = \operatorname{conv}(a, b, d) \cap A$ and points a, b, the conditions of Lemma 13.8 are fulfilled. (Here we use the fact that the linear hull of three points a, b, d may be identified either with $\mathbb{R}^1$ or $\mathbb{R}^2$ (see Theorem 1.21 from [30]).) Hence, by virtue of Lemma 13.8, for all $\alpha \in [0, 1]$ we have $\alpha a + (1 - \alpha)b \in A$. This means that $b \in C$. The obtained contradiction proves the closedness of C. Analogously, it can be proved that the set C is open with respect to the set A. (The latter means that for each point $c \in C$ there exists such an $\varepsilon > 0$, that $O(c, \varepsilon) \bigcap A \subseteq C$.) So, we conclude that $C = A$, since the set A is connected. The theorem is proved. □

Notice, that simple examples demonstrate the fact that in the proved theorem assumptions about connectedness and closedness of the set A are essential: not one of them can be omitted.

Part II: **Set-valued analysis**

14 Introduction to the theory of topological and metric spaces

In this section we will collect the summary of elementary notions and results that we will need in the following. Details can be found in a number of textbooks, for example, in [27, 30, 41].

We will start with general topology. Let X be a set in which a certain family τ of its subsets is selected. It is supposed that this family obeys the following properties: X and the empty set $\varnothing$ belong to τ, the intersection of each two sets from τ also belongs to τ, and lastly, the union of every collection of sets from τ also belongs to τ. The elements of the family τ are called *open sets*. The family τ is called *topology* and the set X with preset topology is called *the topological space* and is denoted by (X, τ).

Notice that different topologies on the same set may exist. If τ_1 and τ_2 are two topologies on X such that $\tau_1 \subset \tau_2$, then one says that the topology τ_1 is weaker than the topology τ_2.

A subset of a topological space is called *closed* if its complement is open. Each open set containing a point x is called its *neighborhood*. A topological space is called *Hausdorff*[1] provided each two of its distinct points have disjoint neighborhoods. In the following we will assume that all considered topological spaces are Hausdorff.

Let (X, τ) be a topological space and A an arbitrary subset of X. Let also O_σ, $\sigma \in \Sigma$ be a family of subsets of X. It is called *the covering of the set A* provided $A \subset \bigcup_{\sigma\in\Sigma} O_\sigma$. If all sets O_σ are open, then this covering is called *open*. A topological space is called *compact* if from each of its open coverings one may select a finite subcovering.

For a subset A of a topological space (X, τ), denote by γ the collection of all intersections $E \bigcap A$, where $E \in \tau$. Then γ is a topology in A, which is called *the induced topology*. A set A is called *compact* if it is a compact set in the induced topology. It is known that each closed subset of a compact set is compact.

A family $\tau' \subset \tau$ is called *the base of the topology τ*, if each element from τ can be represented in the form of the union of elements from τ'. A family $\tau' \subset \tau$ is called *the prebase of the topology τ*, if all possible finite intersections of elements from τ' form its base.

Let (X, τ) and (Y, σ) be two topological spaces and $f : X \to Y$ a map. This map is called *continuous at a point* $x_0 \in X$, if the pre-image of every neighborhood of a point $y_0 = F(x_0)$ contains a certain neighborhood of the point x_0. A mapping is called *continuous on the whole space* (X, τ), if it is continuous at each point $x \in X$. It is easy to verify that a mapping is continuous on the whole space if and only if the pre-image of each open set is open.

---

1 Felix Hausdorff (1868–1942), a German mathematician.

DOI 10.1515/9783110460308-015

It is known that if a topological space (X, τ) has a denumerable base, then, for an arbitrary set $A \subset X$, from every one of its open coverings one may select a finite or denumerable subcovering (see, e.g., [27]).

A typical example of a Hausdorff topological space is *a metric space*. Recall that a metric space is the pair (X, ρ), where X is a certain set and ρ a nonnegative function, called metric, which assigns to each two points $x_1, x_2 \in X$ a number $\rho(x_1, x_2)$ in such a way that the following axioms are fulfilled:
- $\rho(x_1, x_2) = 0 \Leftrightarrow x_1 = x_2$,
- $\rho(x_1, x_2) = \rho(x_2, x_1)\ \forall x_1, x_2 \in X$,
- $\rho(x_1, x_3) \leq \rho(x_1, x_2) + \rho(x_2, x_3)\ \forall x_1, x_2, x_3 \in X$.

The last relation is called *the triangle inequality*. Notice, that in cases where it does not create misunderstandings, we will omit the symbol ρ in the notation of a metric space, denoting it simply by X.

Let (X, ρ) be a metric space and $A \subset X$. Then (A, ρ) is also a metric space which is called *the subspace of X*.

For $\varepsilon > 0$ and $x_0 \in X$, set

$$O^X(x_0, \varepsilon) = \{x \in X : \rho(x, x_0) < \varepsilon\},$$
$$B^X(x_0, \varepsilon) = \{x \in X : \rho(x, x_0) \leq \varepsilon\}.$$

These sets are called, respectively, *open and closed balls of the radius ε with the center at the point* x_0. For an arbitrary set $A \subset X$ and $\varepsilon > 0$, the set $O^X(A, \varepsilon)$, consisting of all such $x \in X$, that $\inf\{\rho(x, a),\ a \in A\} < \varepsilon$ is called *the open ε-neighborhood of A*. Respectively, *the closed ε-neighborhood of the set A* is the set $B^X(A, \varepsilon)$, consisting of all such $x \in X$, that $\inf\{\rho(x, a),\ a \in A\} \leq \varepsilon$. Notice that usually, for convenience, we will omit the upper index denoting the space X in the notation of balls and neighborhoods.

A point x_0 is called *the interior point of a set A*, if there exists an $\varepsilon > 0$ such that $O(x_0, \varepsilon) \subset A$. The set of all interior points of the set A is denoted as $\operatorname{int} A$.

If in a metric space we take the collection τ of all subsets coinciding with their interiors and add to it the empty set $\emptyset$, then it is easy to verify that τ is a topology. Naturally, we can try to introduce any other topologies into a metric space, but in the following, when dealing with metric spaces, we always mean the above mentioned topology, which is sometimes called *the metric topology*. It is clear that a metric space is Hausdorff with respect to this topology.

So, in a metric space a set is open if and only if each of its points is an interior point. It is clear that an open ε-neighborhood of each set is an open set. As in any other topological space, the complement to an open subset of a metric space is called a closed set. It is easy to verify that each closed ε-neighborhood of any set is a closed set.

The intersection of all closed sets containing a set A is called its *closure* and is denoted by *clA*. Obviously, $\operatorname{cl} O(x_0, \varepsilon) \subset B(x_0, \varepsilon)$ but at the same time, there exist metric spaces in which a point x_0 and a number $\varepsilon > 0$ may be found in such a way that $\operatorname{cl} O(x_0, \varepsilon) \neq B(x_0, \varepsilon)$. As an example we may consider the space (X, ρ) with the discrete metric defined by the relation $\rho(x_1, x_2) = 1 \ \forall x_1 \neq x_2$. In this space, for each point x we have $B(x, 1) = X$, $O(x, 1) = \{x\} = \mathrm{cl}\, O(x, 1)$, and every set is open.

A widely known example of a metric space is an arbitrary subset of a normed space endowed with the metric $\rho(x_1, x_2) = \|x_1 - x_2\|$. It is a simple exercise to verify that all the properties of a metric are satisfied for this function.

Let (X, ρ) be a metric space. We will say that a sequence of points $\{x_n\} \subset X$ *converges to a point* $x_0 \in X$ (or *has a limit* x_0), if $\rho(x_n, x_0) \to 0$, $n \to \infty$. A point $x_0 \in X$ is called *a limit point of a set* $C \subset X$, if each neighborhood of a point x_0 contains a point $x \in C\colon x \neq x_0$. It is known that a set C is closed if and only if it contains all its limit points.

Similarly to normed spaces, a metric space is called *separable* if it contains a denumerable everywhere dense subset $\{x_1, x_2, \dots, x_n, \dots\}$. The latter means that for arbitrary $x \in X$ and $\varepsilon > 0$ there exists such a number $n = n(x, \varepsilon)$, that $\rho(x_n, x) \leq \varepsilon$.

It is easy to prove (see, e.g., [27]), that a metric space is separable if and only if the topology induced by its metric has a denumerable base. From here it follows that an arbitrary subspace of a separable metric space is separable itself (with respect to the induced topology). Indeed, it is clear that the induced topology also possesses a denumerable base and hence the subspace is separable.

A sequence $\{x_n\} \subset X$ is called *fundamental* if for an arbitrary $\varepsilon > 0$ there exists such a number $N(\varepsilon)$, that $\rho(x_n, x_m) \leq \varepsilon \ \forall n, m \geq N(\varepsilon)$. It is easy to see that each convergent sequence is fundamental. If in a metric space every fundamental sequence is convergent, it is called *complete*.

As examples of complete metric separable spaces we can consider: the space $\mathbb{R}^n$ with the natural metric $\rho(x_1, x_2) = |x_2 - x_1|$, the space $C[a, b]$ of continuous real-valued functions $x(t)$, $t \in [a, b]$ with the metric defined by the formula $\rho(x_1, x_2) = \max_{t \in [a,b]} |x_2(t) - x_1(t)|$, the Hilbert space l_2 and others.

Consider criteria of compactness in metric spaces. Let (X, ρ) be a metric space and $C \subset X$. A set C is compact if and only if from each sequence of its points $\{x_n\} \subset C$ we may select a subsequence converging to a point $x_0 \in C$.

Further, for $\varepsilon > 0$ a set of points C' is called an ε-net of a set C, provided an ε-neighborhood of C' contains the set C. A set is called *totally bounded* (or *precompact*) if for an arbitrary $\varepsilon > 0$ it has a finite (i.e., consisting of a finite number of elements) ε-net. It is known that a set is compact if and only if it is totally bounded and complete (i.e., each fundamental sequence lying in it converges to a point of this set).

Let (X, ρ) and (Y, d) be two metric spaces and $f\colon X \to Y$ a map. This map is called *continuous at a point* $x_0 \in X$, if for each sequence $\{x_n\}$, which converges to x_0, we have $f(x_n) \to f(x_0)$, $n \to \infty$. A map is called *continuous* (*on the whole space* (X, ρ)) provided it is continuous at each point $x \in X$. The introduced definition of continuity of maps in metric spaces is often called *sequential*, since it is formulated "in the language" of sequences. It is easy to see that it is equivalent to the following definition of the continuity of f at a point x_0 which is in accordance with the general definition of continuity given above: for each $\varepsilon > 0$ there exists such a $\delta > 0$ that $f(O^X(x_0, \delta)) \subset O^Y(f(x_0), \varepsilon)$.

15 The Hausdorff metric and the distance between sets

Let (X, ρ) be a metric space. Remember that a set is called bounded if it is contained in a certain ball. It is clear that each totally bounded and, moreover, compact set is bounded (but, naturally, not vice versa). It is also evident that every subset of a bounded set is bounded.

Consider the notion of *the Hausdorff distance* between two nonempty bounded sets. So, let $M, N \subseteq X$ be nonempty bounded sets. The Hausdorff distance between M and N is defined as

$$h(M, N) = \inf\{ r > 0 \colon O(M, r) \supseteq N,\ O(N, r) \supseteq M \}. \tag{15.1}$$

Notice that the existence of corresponding finite $r > 0$ follows from the boundedness of sets M, N. Moreover, if in the above formula we replace open neighborhoods with closed ones, the value of infimum in it will be the same.

It is evident that for each $x \in M$ and $\varepsilon > 0$ there exists $y \in N$ such that $\rho(x, y) \leq h(M, N) + \varepsilon$. Furthermore, if each of the sets M, N is a singleton (i.e., consists of a single point), then $h(M, N) = \rho(M, N)$.

Let us mention that the above formula defines also the Hausdorff distance between unbounded sets M and N, if we allow $h(M, N)$ to take the value $+\infty$.

Present an example of the evaluation of the Hausdorff distance between two balls in a metric space.

Proposition 15.1. *For each $a_1, a_2 \in X$ and $r_1, r_2 > 0$ the following inequality holds*

$$h(B(a_1, r_1), B(a_2, r_2)) \leq \rho(a_1, a_2) + \max\{ r_1, r_2 \}. \tag{15.2}$$

If we additionally assume that X is a linear normed space then

$$h(B(a_1, r_1), B(a_2, r_2)) = \|a_2 - a_1\| + |r_2 - r_1|. \tag{15.3}$$

Proof. Prove (15.2). For an arbitrary $x_1 \in B(a_1, r_1)$ we have

$$\rho(x_1, a_2) \leq r_1 + \rho(a_1, a_2). \tag{15.4}$$

As the result of the evident chain of inequalities we get

$$\rho(x_1, a_2) \leq \rho(x_1, a_1) + \rho(a_1, a_2) \leq r_1 + \rho(a_1, a_2).$$

Similarly it can be proved that $\rho(x_2, a_1) \leq r_2 + \rho(a_1, a_2)$ for all $x_2 \in B(a_2, r_2)$. From the obtained inequalities (15.2) immediately follows.

DOI 10.1515/9783110460308-016

Now, let X be a normed space. Set $\kappa = \|a_2 - a_1\| + |r_2 - r_1|$. Let us prove that $h(B(a_1, r_1), B(a_2, r_2)) \le \kappa$. Take an arbitrary $x_1 \in B(a_1, r_1)$. If $\|a_2 - x_1\| > r_2$, then set $\xi = a_2 - (a_2 - x_1)\|a_2 - x_1\|^{-1} r_2$. Then $\xi \in B(a_2, r_2)$ and by (15.4) we get $\|\xi - x_1\| \le \|a_2 - x_1\| - r_2 \le \kappa$. In case $\|a_2 - x_1\| \le r_2$, we have $x_1 \in B(a_2, r_2)$, and we set $\xi = x_1$. So, we proved the existence of $\xi \in B(a_2, r_2)$ such that $\|\xi - x_1\| \le \kappa$. Similarly we can prove that for each $x_2 \in B(a_2, r_2)$ there exists $\xi \in B(a_1, r_1)$, for which $\|\xi - x_2\| \le \kappa$ and hence $h(B(a_1, r_1), B(a_2, r_2)) \le \kappa$.

Let us prove the inverse inequality. We will assume that $a_1 \ne a_2$ (otherwise the inequality is evident). For the determinacy, set $r_1 \ge r_2$. Take $x_1 = a_1 + (a_1 - a_2)\|a_2 - a_1\|^{-1} r_1$. Then $x_1 \in B(a_1, r_1)$, and, by the triangle inequality, for an arbitrary $x \in B(a_2, r_2)$ we get

$$\begin{aligned} \|x - x_1\| &\ge \|x_1 - a_2\| - \|a_2 - x\| \ge \|x_1 - a_2\| - r_2 \\ &= \|a_2 - a_1\| + r_1 - r_2 = \kappa, \end{aligned}$$

implying $h(B(a_1, r_1), B(a_2, r_2)) \ge \kappa$, that proves (15.3). □

The following example demonstrates that inequality (15.2) may turn into the equality.

Example 15.2. Consider the metric space (X, ρ), where $X = \{-2, -1, 1, 2\}$ is the set consisting of four points on the real axis endowed with the natural metric. Then for $a_1 = -1$, $a_2 = 1$ and $r_1 = r_2 = 1$ we have

$$h(B(a_1, r_1), B(a_2, r_2)) = 3 = \rho(a_1, a_2) + \max\{r_1, r_2\}.$$

It is easy to see that for closed bounded subsets, the Hausdorff distance satisfies all axioms of a metric. Namely, for each nonempty closed bounded sets M, N, and E we have

$$\begin{gathered} h(M, N) = 0 \Leftrightarrow M = N, \quad h(M, N) = h(N, M) \ge 0, \\ h(M, N) \le h(M, E) + h(E, N). \end{gathered}$$

Therefore, the function h turns the set of all nonempty closed bounded subsets of the space X into the metric space, which we will denote by $\mathcal{H}(X)$ (indicating in this notation neither the metric ρ nor the metric h). The metric h is called *the Hausdorff metric.*

Along with the metric space $\mathcal{H}(X)$ we will consider its subspace $\mathcal{H}_c(X)$, *consisting of nonempty compact subsets of* X. It is clear that $\mathcal{H}_c(X)$, endowed with the Hausdorff metric h, is also the metric space. Notice that if the set X is bounded, then the function h is bounded on the set of all subsets of X, i.e., there exists such an $r > 0$, that $h(M, N) \le r$ for each $M, N \subset X$.

Theorem 15.3. *Let a metric space X is separable. Then the metric space $\mathcal{H}_c(X)$ is also separable.*

Proof. Choose in X a denumerable everywhere dense subset $A \subset X$. Denote by $\mathcal{A}$ the family of all possible finite subsets of A. It is known that the family $\mathcal{A}$ is denumerable.

Let us show that it is everywhere dense in the space $\mathcal{H}_c(X)$. Indeed, let C be a nonempty compact subset of X and $\varepsilon > 0$ a given number. Take in C a finite $\varepsilon/2$-net, formed by points $c_1, \dots, c_m \in C$. Choose such points $a_i \in A$, that $\rho(a_i, c_i) < \varepsilon/2$, $i = \overline{1, m}$. Then, obviously

$$O(\{a_1, \dots, a_m\}, \varepsilon) = \bigcup_{i=1}^{m} O(a_i, \varepsilon) \supset \bigcup_{i=1}^{m} O(c_i, \varepsilon/2) \supset C,$$

$$O(C, \varepsilon) \supset \bigcup_{i=1}^{m} O(c_i, \varepsilon) \supset \{a_1, \dots, a_m\}.$$

Therefore $h(\{a_1, \dots, a_m\}, C) \leq \varepsilon$. So, we proved that the denumerable family $\mathcal{A}$ is everywhere dense in the space $\mathcal{H}_c(X)$. □

The following assertion shows that in the case when the set X is totally bounded, the space $\mathcal{H}_c(X)$ in Theorem 15.3 may be replaced with the "larger" space $\mathcal{H}(X)$.

Theorem 15.4. *Let a metric space X be totally bounded. Then the metric space $\mathcal{H}(X)$ is separable.*

Proof. First let us show that the space X is itself separable. Indeed, for each natural number n, take a finite $1/n$-net in a totally bounded space X and consider the union with respect to all natural n of points forming these finite nets. The constructed set A is denumerable and, as is easy to see, everywhere dense in X.

Define the family $\mathcal{A}$ as during the proof of Theorem 15.3. Take an arbitrary nonempty subset $C \subset X$. It is totally bounded as a subset of the totally bounded set X and, hence, for each $\varepsilon > 0$ it has a finite ε-net. Repeating for the set C the same reasoning as during the proof of Theorem 15.3, we get the desired result. □

In connection with the proved theorem, notice that if a metric space X is totally bounded, but noncompact, then $\mathcal{H}(X) \neq \mathcal{H}_c(X)$, since $X \in \mathcal{H}(X)$, however $X \notin \mathcal{H}_c(X)$.

The following example firstly shows that the space $\mathcal{H}_c(X)$ in Theorem 15.3 cannot be replaced, generally speaking, with a "larger" space $\mathcal{H}(X)$, and, secondly, it is also impossible to weaken the assumptions in Theorem 15.4 replacing the condition of total boundedness of the space X with its boundedness and separability.

Example 15.5. As a metric space (X, ρ) take the closed unit ball $B_1 = \{x \in l_2 : |x| \leq 1\}$ in a Hilbert space l_2 with the natural metric generated by the norm. The set X is obviously bounded (but not totally bounded) and separable (see, e.g., [27]). There exists a denumerable set A of such points $\{a_i\}$ from X (they form the orthonormal basis in l_2), that $\rho(a_i, a_j) \geq \sqrt{2}$ $\forall i \neq j$. Denote by $\mathcal{A}$ the family of all subsets of A. Then $\mathcal{A}$ has the continuum power and, as is easy to see, $h(A_1, A_2) \geq \sqrt{2}$ $\forall A_1, A_2 \in \mathcal{A} : A_1 \neq A_2$. So, in the space $\mathcal{H}(X)$ there exists a continual family of elements $\mathcal{A}$ such that the Hausdorff distance between each two of its elements is no less than $\sqrt{2}$. This easily yields (prove it as an exercise!) that the space $\mathcal{H}(X)$ is not separable.

Theorem 15.6. *Let a metric space X be complete. Then both spaces $\mathcal{H}(X)$ and $\mathcal{H}_c(X)$ are complete.*

Proof. First let us prove, following [13], § 1.3, the completeness of the space $\mathcal{H}(X)$. Let $\{A_n\}$ be a fundamental sequence in $\mathcal{H}(X)$. It is necessary to find such a nonempty closed bounded set $A \subset X$ that $h(A, A_n) \to 0$, $n \to \infty$.

Choose a subsequence $\{A_{n_k}\}$ by condition

$$h(A_{n_k}, A_{n_{k+1}}) < 1/2^{k+1}, \quad k = 1, 2, \dots, \tag{15.5}$$

and set

$$A = \bigcap_{k=1}^{\infty} B(A_{n_k}, 2^{-k}). \tag{15.6}$$

The set A is closed because it is the intersection of closed sets and bounded since it is contained in the bounded set $B(A_{n_1}, 1/2)$. Now it is sufficient to show that the set A is nonempty and $h(A, A_{n_k}) \to 0$, $k \to \infty$. Indeed, if A is nonempty, then by the triangle inequality we have $h(A_n, A) \le h(A_{n_k}, A_n) + h(A_{n_k}, A)$. This yields, by the fundamentality of the sequence $\{A_n\}$, that $h(A, A_n) \to 0$, $n \to \infty$.

Fix a number k and take an arbitrary point $y \in A_{n_k}$. By (15.5), there exists such a sequence of points $\{x_{n_m}\}$, that $x_{n_k} = y$, $x_{n_m} \in A_{n_m}$ and $\rho(x_{n_m}, x_{n_{m+1}}) \le 1/2^{m+1}$ for all $m \ge k$. For each m and $s > m$ we have

$$\rho(x_{n_s}, x_{n_m}) \le \sum_{l=m}^{s-1} \rho(x_{n_l}, x_{n_{l+1}}) \le \sum_{l=m}^{s-1} 2^{-(l+1)} \le 2^{-m}.$$

So, the sequence $\{x_{n_m}\}$ is fundamental and hence, by the completeness of X, there exists its limit $x = \lim_{m\to\infty} x_{n_m}$. Furthermore, $\rho(x_{n_m}, x) \le 1/2^m$ $\forall m \ge k$. Therefore for each $m \ge k$ we get $x \in B(x_{n_m}, 2^{-m}) \subset B(A_{n_m}, 2^{-m})$. For $m = 1, \dots, k-1$, by virtue of (15.5) we have

$$x \in B(A_{n_k}, 2^{-k}) \subset B(A_{n_{k-1}}, 2^{-(k-1)}) \subset \dots \subset B(A_{n_m}, 2^{-m})$$

and hence, $x \in A$, justifying, in particular, that the set A is nonempty.

Since the point $y \in A_{n_k}$ was taken arbitrarily, we get the inclusion

$$A_{n_k} \subset B(A, 2^{-k}) \quad \forall k.$$

By (15.6), the inclusions $A \subset B(A_{n_k}, 2^{-k})$ $\forall k$ are true. From the obtained inclusions it follows that $h(A_{n_k}, A) \le 1/2^k$ and hence $h(A_{n_k}, A) \to 0$, $k \to \infty$.

Let us prove now that the space $\mathcal{H}_c(X)$ is complete. Let $\{A_n\}$ be a fundamental sequence of elements from the space $\mathcal{H}_c(X)$. Choose a subsequence A_{n_k} by condition (15.5) and define the set A by relation (15.6). By virtue of the above reasoning it is sufficient to prove that the set A is compact. Let us do it.

As mentioned above, the set A is closed. Let us show that it is totally bounded. Indeed, take an arbitrary $\varepsilon > 0$. Choose a number k in such a way that $2^{-k} < \varepsilon/2$. Then by virtue of (15.6) we get $B(A_{n_k}, \varepsilon/2) \supseteq A$. The set A_{n_k} is compact. So, it has a finite $\varepsilon/2$-net $a_1, \dots, a_m$. Hence,

$$\bigcup_{i=1}^{m} B(a_i, \varepsilon) \supseteq B(A_{n_k}, \varepsilon/2) \supseteq A.$$

Therefore, A is totally bounded. But a closed totally bounded subset of a complete metric space is compact and hence the set A is compact. □

Another proof of Theorem 15.6 can be found in [41], ch. 4, § 7.

If a metric space X is compact, then it is complete and $\mathcal{H}(X) = \mathcal{H}_c(X)$. Therefore, in this case the validity of Theorem 15.6 may be deduced from the following nontrivial assertion.

Theorem 15.7. *Let a metric space X be compact. Then the space $\mathcal{H}(X)$ is also compact.*

The proof of this theorem can be found, for example, in [13], § 1.3.

Let subsets $M, N \subset X$ be given. Along with the Hausdorff distance, another type of distance between these sets is used. This distance will be denoted by dist, it is defined by the relation

$$\operatorname{dist}(M, N) = \inf\{\rho(x, y),\ x \in M,\ y \in N\}.$$

For so defined distance we obviously will have

$$\operatorname{dist}(M, N) \le h(M, N), \quad M \bigcap N \ne \emptyset \Rightarrow \operatorname{dist}(M, N) = 0,$$
$$\operatorname{dist}(M, N) = \operatorname{dist}(N, M) \ge 0.$$

Consider the distance between two balls with their centers at the points $a_1, a_2 \in X$ and of radii r_1 and r_2 respectively. In an arbitrary metric space, evidently, the following estimate holds true

$$\operatorname{dist}(B(a_1, r_1), B(a_2, r_2)) \le \rho(a_1, a_2).$$

If we assume, additionally, that X is a linear normed space then we get

$$\operatorname{dist}(B(a_1, r_1), B(a_2, r_2)) = \max(0, \|a_1 - a_2\| - (r_1 + r_2)). \tag{15.7}$$

Exercise 15.8. *Verify this equality and compare it with formulae* (15.2) *and* (15.3) *obtained for the Hausdorff distance.*

Let us show that if sets M and N are closed, at least one of them is compact and $\operatorname{dist}(M, N) = 0$, then the intersection of these sets is nonempty. Indeed, let, for determinacy, the set M be compact. By virtue of $\operatorname{dist}(M, N) = 0$ there exist such sequences $\{x_n\} \subset M$, $\{y_n\} \subset N$, that $\rho(x_n, y_n) \to 0$, $n \to \infty$. By compactness of M, passing to subsequences, we may assume without loss of generality that $x_n \to x_0 \in M$. But then by virtue of $\rho(x_n, y_n) \to 0$ we get $y_n \to x_0$, implying $x_0 \in N$, since the set N is closed. Therefore we have that $x_0 \in M \bigcap N$.

Notice that the assumption about the compactness of at least one of the sets M or N is essential. Indeed, if M and N are the graphs of hyperbolas on the plane defined by the equations $x^1x^2 = 1$ and $x^1x^2 = -1$ respectively, then $\operatorname{dist}(M, N) = 0$, however these sets are disjoint.

The difference between the Hausdorff distance and the distance dist can be characterized by the following example. Let $M = B((0, 0), 1)$, $N = B((3, 0), 1)$ be unit discs on the plane. Then, in accordance with (15.3) and (15.7) we get $h(M, N) = 3$, $\operatorname{dist}(M, N) = 1$. Moreover, this example shows that sets M and N can contain points such that the distance between them is greater than the Hausdorff distance between M and N: the distance between $(-1, 0) \in M$ and $(4, 0) \in N$ equals 5.

Introduce into consideration one more quantity characterizing the mutual location of two subsets M and N of a space X. This is *the deviation of the set M from the set N*, which is defined by the relation

$$h^+(M, N) = \inf\{\, \varepsilon > 0 : O(N, \varepsilon) \supset M \,\}.$$

In the case when $\varepsilon > 0$ satisfying $O(N, \varepsilon) \supset M$ does not exist we set, by definition, $h^+(M, N) = \infty$. But it is clear that if the set M is bounded, the value of $h^+(M, N)$ is finite.

Exercise 15.9. *Prove that the deviation h^+ may be equivalently defined as*

$$h^+(M, N) = \sup\{\, \operatorname{dist}(x, N),\ x \in M \,\},$$

and the Hausdorff distance h may be expressed through the deviation h^+ by the formula

$$h(M, N) = \max\{\, h^+(M, N), h^+(N, M) \,\}.$$

So, for each sets M, N we have

$$\operatorname{dist}(M, N) \le h^+(M, N) \le h(M, N).$$

It is clear that if $M \subset N$, then $h^+(M, N) = 0$, but if, for example $N = \{\, a, b \,\}$ and $M = \{\, a \,\}$ then $h^+(N, M) > 0$. It means that the arguments of the function h^+ cannot be interchanged, i.e., generally speaking, $h^+(M, N) \ne h^+(N, M)$.

Exercise 15.10. *Prove that the deviation h^+ satisfies the triangle inequality: for arbitrary subsets E, M, N of X the following inequality holds true:*

$$h^+(E, N) \le h^+(E, M) + h^+(M, N).$$

Let us return to the properties of the distance dist. Contrary to the Hausdorff distance h and the deviation h^+, the triangle inequality for the distance dist, generally speaking, does not hold. As confirmation, we can consider the example of two nonempty disjoint compact sets E, N and the set $M = E \bigcup N$. Then $E \bigcap M \neq \emptyset$, $M \bigcap N \neq \emptyset$, but $E \bigcap N = \emptyset$. So, obviously, $\operatorname{dist}(E, N) > 0$, however $\operatorname{dist}(E, M) = \operatorname{dist}(M, N) = 0$.

At the same time, for arbitrary sets E, M, N from X we have

$$\operatorname{dist}(E, M) \leq \operatorname{dist}(E, N) + h^+(N, M). \tag{15.8}$$

Let us prove this inequality. Take an arbitrary $\varepsilon > 0$. There exist such $n \in N$, $e \in E$, that $\rho(n, e) \leq \operatorname{dist}(E, N) + \varepsilon$. By definition of the deviation of the set N from the set M there exists such an $m \in M$, that $\rho(m, n) \leq h^+(N, M) + \varepsilon$. By the triangle inequality we get $\rho(m, e) \leq \rho(m, n) + \rho(n, e)$. Furthermore, $\operatorname{dist}(M, E) \leq \rho(m, e)$. From the obtained inequalities we have $\operatorname{dist}(E, M) \leq \operatorname{dist}(E, N) + h^+(N, M) + 2\varepsilon$. From here, by the arbitrariness of $\varepsilon > 0$ we get (15.8).

Inequality (15.8) yields the following. If both of the sets M and N are singletons: $M = \{x_1\}$, $N = \{x_2\}$, then $h^+(M, N) = \rho(x_1, x_2)$ and, by (15.8) we have the following triangle type inequality:

$$\operatorname{dist}(E, \{x_1\}) \leq \operatorname{dist}(E, \{x_2\}) + \rho(x_1, x_2). \tag{15.9}$$

Let us discuss inequality (15.8). It is nonsymmetric in the following sense. In its left-hand side we have the expression $\operatorname{dist}(E, M)$, which does not vary after the interchange of variables E and M, i.e., $\operatorname{dist}(E, M) = \operatorname{dist}(M, E)$. At the same time, after such an interchange of E and M the expression in the right-hand side of (15.8) may vary. This leads us to the idea of receiving the corresponding symmetric estimate.

Let us do it. Interchanging the sets E and M in inequality (15.8) and using the equality $\operatorname{dist}(E, M) = \operatorname{dist}(M, E)$, we get

$$\operatorname{dist}(E, M) \leq \operatorname{dist}(M, N) + h^+(N, E).$$

From here and (15.8) immediately follow the desired inequalities

$$\operatorname{dist}(E, M) \leq \max\left(\operatorname{dist}(E, N), \operatorname{dist}(M, N)\right) + \min\left(h^+(N, M), h^+(N, E)\right),$$
$$\operatorname{dist}(E, M) \leq \min\left(\operatorname{dist}(E, N), \operatorname{dist}(M, N)\right) + \max\left(h^+(N, M), h^+(N, E)\right).$$

It is easy to see that inequality (15.8) is the corollary of each of the obtained inequalities. At the same time, every one of these inequalities is not the consequence of the other one, i.e., there exist examples of sets (find them as an exercise) for which the right-hand side of the first inequality is greater than the right-hand side of the second inequality and vice versa.

16 Some fine properties of the Hausdorff metric

The definition of the Hausdorff distance gives rise to the following natural question. Let M, N be two nonempty closed bounded subsets of a metric space (X, ρ) and $h(M, N) = r$. Then, according to (15.1) for each $x \in M$ and $\varepsilon > 0$ there exists $y \in N$ such that $\rho(x, y) \leq r + \varepsilon$. The question is, does a pair (x, y) exist such that $x \in M$, $y \in N$ and $\rho(x, y) \leq r$?

In this section we will answer this question.[1] First we will prove that under an additional assumption, known as the Bolzano–Weierstrass condition, for each $x \in M$ there exists $y \in N$ such that $\rho(x, y) \leq h(M, N)$. Then we will show in Theorem 16.9 that in general the answer is negative. More precisely, we will prove that in each infinite-dimensional normed space there exist closed bounded sets M, N such that $\|x - y\| > h(M, N)$ for all $x \in M$, $y \in N$. The results of this section are based on the paper [42].

16.1 Hausdorff distance between sets satisfying the Bolzano–Weierstrass condition

Let us recall that a set satisfies *the Bolzano–Weierstrass condition* if each its bounded sequences contains a convergent subsequence.

Theorem 16.1. *Let M and N be closed sets and let the set N satisfy the Bolzano–Weierstrass condition. Then*

$$\forall x \in M \; \exists y \in N\colon \rho(x, y) \leq h(M, N). \tag{16.1}$$

Proof. Set $r = h(M, N)$. Without loss of generality assume that $r < \infty$. Let $x \in M$. The definition of the Hausdorff distance implies that for each natural n there exists $y_n \in N$ such that $\rho(y_n, x) \leq r + 1/n$. So, the sequence $\{y_n\}$ is bounded. Therefore, there exist a subsequence $\{y_{n_m}\}$ and a point $y \in X$ such that $\{y_{n_m}\}$ converges to y. By assumption, N is closed, that means that $y \in N$. Finally, passing to the limit as $m \to \infty$ in the inequality $\rho(y_{n_m}, x) \leq r + 1/n_m$ we get $\rho(x, y) \leq r$. □

Obviously, every compact set satisfies the Bolzano–Weierstrass condition. Moreover, if $X = \mathbb{R}^n$, then the classical Bolzano–Weierstrass theorem yields that every closed set $N \subset \mathbb{R}^n$ satisfies the Bolzano–Weierstrass condition.

Remark 16.2. The converse of this implication is not true. More precisely, there exists a space with an infinite number of elements such that condition (16.1) is satisfied for every closed set M, N; however, it possesses a bounded sequence without limit points.

1 During an initial reading of this book, this section may be omitted.

DOI 10.1515/9783110460308-017

Indeed, let X be an infinite set, ρ be a discrete metric on X, i.e.,

$$\rho(x, y) = \begin{cases} 0, & x = y, \\ 1, & x \neq y. \end{cases}$$

Obviously, all subsets of X are closed. Furthermore, evidently $h(M, N) = 1$ if and only if $M \neq N$. So, (16.1) is true. At the same time, let $\{x_n\}$ be a sequence such that $x_n \neq x_m$ for all $n \neq m$. It is obviously bounded, but it has no limit points.

Remark 16.3. Under the assumptions of Theorem 16.1 the inequality in (16.1) cannot be replaced with the corresponding equality. Indeed, in the following example all assumptions of Theorem 16.1 are satisfied, however,

$$\forall x \in M \ \forall y \in N: \rho(x, y) < h(M, N).$$

Let $X = l_2$, $N = \{0\}$,

$$M = \{x_n = (x_1^n, x_2^n, \dots): n = 1, 2, \dots, \ x_n^n = 1 - n^{-1}, \ x_j^n = 0 \ \forall j \neq n\}.$$

Then $h(M, N) = 1$, however, $\rho(0, x_n) = 1 - n^{-1} < 1$ for each n.

Let us present one more example in which X satisfies the Bolzano–Weierstrass condition, however,

$$\forall x \in M \ \forall y \in N: \rho(x, y) \neq h(M, N)$$

and there exist points $x, u \in M$, $y, v \in N$ such that

$$\rho(x, y) < h(M, N), \quad \rho(u, v) > h(M, N).$$

Let $X = \mathbb{R}$,

$$M = \{x_n = 3n: n = 2, 3, \dots\}, \quad N = \{y_n = 3n - 1 + n^{-1}: n = 2, 3, \dots\}.$$

Then for all $n \geq 2$ and $j \geq 2$ we have $\rho(x_n, y_j) = |3(n - j) + 1 - j^{-1}| \neq 1$. However, $\rho(x_n, y_n) = 1 - n^{-1} < 1$ for all n, and $\rho(x_2, y_3) = 2 + 3^{-1} > 1$.

The following assertion shows that in the case of compact sets, there exists a couple of their points on which the Hausdorff distance between these sets is “realized”.

Proposition 16.4. *Let sets M and N be compact. Then*

$$\exists x \in M \ \exists y \in N: \rho(x, y) = h(M, N).$$

Proof. The definition of the Hausdorff distance yields

$$h(M, N) = \max\{\sup_{x\in M}(\inf_{y\in N}\rho(x, y)), \sup_{y\in N}(\inf_{x\in M}\rho(x, y))\}.$$

Without loss of generality we can assume that

$$h(M, N) = \sup_{x\in M}(\inf_{y\in N}\rho(x, y)).$$

It is not difficult to verify (you can do it as an exercise!) that the function $x \mapsto \inf_{y \in N} \rho(x, y)$ is continuous. Since the set M is compact, there exists a point $x \in M$ such that $\inf_{y \in N} \rho(x, y) = h(M, N)$. Similarly, the continuity of the function $y \mapsto \rho(x, y)$ and the compactness of the set N yields the existence of a point $y \in N$ such that $\rho(x, y) = h(M, N)$. □

Notice that if the set M is bounded and $h(M, N) < \infty$, then the set N is bounded. It is obvious that in this case the set N is compact provided it satisfies the Bolzano–Weierstrass condition.

16.2 Hausdorff distance between subsets of normed spaces

First let us recall the definition of the orthogonality in normed spaces and prove some auxiliary propositions.

Definition 16.5. A vector x is called orthogonal to a vector y in a normed space X if

$$\|x\| \leq \|x + \alpha y\| \quad \forall \alpha.$$

We will denote this by $x \perp y$. A vector x is called orthogonal to a subspace $L \subset X$ if

$$x \perp y \quad \forall y \in L,$$

that is equivalent to the following relation:

$$\|x\| \leq \|x + y\| \quad \forall y \in L. \tag{16.2}$$

Definition 16.6. The sequence of vectors $\{e_n\}_{n=1}^{\infty}$ in a normed space X is called orthonormal if for all positive integers n we have

$$\|e_n\| = 1, \quad e_{n+1} \perp L(e_1, \dots, e_n), \tag{16.3}$$

where $L(e_1, \dots, e_n)$ denotes a linear hull of the vectors $e_1, \dots, e_n$.

Lemma 16.7. *Let X be a finite-dimensional space and L be a subspace such that $L \neq X$. Then there exists $x \in X$ such that $x \neq 0$ and $x \perp L$.*

Proof. Since $L \neq X$, there exists $y \in X$ such that $y \notin L$. Define a function $g\colon L \to \mathbb{R}$ by the formula $g(v) = \|v + y\|$, for all $v \in L$. Obviously, g is continuous and

$$\exists R \geq 0: \quad \inf_{v \in B(0,R) \cap L} g(v) = \inf_{v \in L} g(v).$$

So, the Weierstrass theorem implies that there exists $v_0 \in L$ such that

$$g(v_0) = \inf_{v \in L} g(v). \tag{16.4}$$

Set $x = y + v_0$. It follows from (16.4) that

$$\|x\| = g(v_0) \leq g(v_0 + v) = \|y + v_0 + v\| = \|x + v\|$$

for all $v \in L$. Moreover, $x \neq 0$. Otherwise $y = -v_0 \in L$, that contradicts the assumption that $y \notin L$. □

Lemma 16.8. *In every infinite-dimensional normed space there exists an orthonormal (in the sense of* (16.3)*) sequence.*

Proof. Let X be an infinite-dimensional normed space. We will carry out the proof by induction on k. First let us choose a unit vector $e_1 \in X$. By the infinite-dimensionality of X there exists a vector $f_2 \in X$ such that the vectors e_1, f_2 are linearly independent. So, $L(e_1)$ is a proper subspace of the finite-dimensional space $L(e_1, f_2)$. According to Lemma 16.7 there exists a unit vector $e_2 \in L(e_1, f_2)$ such that $e_2 \perp e_1$. Assume now that the unit vectors $e_1, \dots, e_k$ such that $e_i \perp L(e_1, \dots, e_{i-1})$ for all $i = \overline{2, k}$ are already constructed. Since X is infinite-dimensional there exists vector f_{k+1} such that vectors $e_1, \dots, e_k, f_{k+1}$ are linearly independent. So, $L(e_1, \dots, e_k)$ is a proper space of the finite-dimensional space $L(e_1, \dots, e_k, f_{k+1})$. According to Lemma 16.7 there exists a unit vector $e_{k+1} \in L(e_1, \dots, e_k, f_{k+1})$ such that $e_{k+1} \perp L(e_1, \dots, e_k)$. So, the desired orthonormal sequence exists. □

Let us now present a statement that gives the negative answer to the question raised at the beginning of this section.

Theorem 16.9. *In every infinite-dimensional normed space X there exist nonempty bounded closed sets N and M such that*

$$\|x - y\| > h(M, N)$$

for all $x \in M$ and $y \in N$.

Proof. Let us construct the sets M and N. According to Lemma 16.8 there exists an orthonormal sequence $\{e_n\}_{n=1}^{\infty}$. Set $f_n = \frac{n+1}{n} e_n$. Then

$$\left\| f_n - \sum_{k=1}^{n-1} \alpha_k f_k \right\| \geq \|f_n\| = 1 + \frac{1}{n}. \tag{16.5}$$

Denote by M the set of all the vectors $x \in X$ that can be represented as a sum of an even amount of vectors $f_1, f_2, \dots$. Every vector $x \in M$ can be uniquely represented in such a form because $f_1, f_2, \dots$ are linearly independent. For all $x \in M$ denote by $m(x)$ the greatest number of the corresponding summands.

Denote by N the set of all vectors $y \in X$ that can be represented as a sum of an odd amount of vectors $f_1, f_2, \dots$. Every vector $y \in N$ can be uniquely represented in such a form. For every $y \in N$ denote by $n(y)$ the greatest number of the corresponding summands.

By inequality (16.5) we have that M and N are closed. Furthermore, we have

$$\|y - x\| \geq 1 + \frac{1}{\max\{\,n(x), m(y)\,\}} > 1$$

for arbitrary $x \in M$, $y \in N$. Now let us estimate the Hausdorff distance $h(M, N)$. For all $x \in M$ the vector $y = x + f_k$ belongs to the set N for every $k > m(x)$. Moreover, $\|y - x\| = \|f_k\| \to 1$ as $k \to \infty$. In a similar manner for all $y \in N$ the vector $x = y + f_k$ belongs to the set M for every $k > n(y)$. Further, $\|x - y\| = \|f_k\| \to 1$ when $k \to \infty$. Hence, $h(M, N) \leq 1$. Thus, we constructed closed bounded sets $M \subset X$ and $N \subset X$ such that $h(M, N) \leq 1$ and $\|x - y\| > 1$ for all $x \in M$, $y \in N$. □

Summing up, we get the following characterization of finite-dimensional normed spaces.

Proposition 16.10. *A normed space X is finite-dimensional if and only if for all non-empty closed subsets $M \subset X$ and $N \subset X$ the following relation holds:*

$$\forall x \in M \; \exists y \in N\colon \; \|x - y\| \leq h(M, N).$$

17 Set-valued maps. Upper semicontinuous and lower semicontinuous set-valued maps

Let (X, ρ_X) and (Y, ρ_Y) be metric spaces. Consider a map F which assigns to each $x \in X$ a nonempty closed subset $F(x) \subset Y$. Such a map will be called *set-valued*. Set-valued maps differ from the classical ones (which we will sometimes call "*single-valued*") since each set-valued map assigns to a point x not a single point, but a whole set. Therefore, the collection of set-valued maps naturally includes classical maps and set-valued maps may be considered essential generalizations of usual ones.

A set-valued map F is called *compact-valued* provided its values $F(x)$ are compact for each $x \in X$. In the case where Y is a normed space, a set-valued map F is called *convex-valued* if all the values $F(x)$ are convex sets.

Let us describe the following important classes of set-valued maps.

Definition 17.1. A set-valued map F is called sequentially upper semicontinuous at a point $x_0 \in X$, if for each sequence $\{x_n\}$, convergent to the point x_0, and every sequence $\{y_n\}$, such that $y_n \in F(x_n)$ $\forall n$, we have $\operatorname{dist}(y_n, F(x_0)) \to 0$, $n \to \infty$.

If a set-valued map is sequentially upper semicontinuous at each point it is called *sequentially upper semicontinuous*.

Definition 17.2. A set-valued map F is called sequentially lower semicontinuous at a point $x_0 \in X$, if for each sequence $\{x_n\}$, convergent to the point x_0, and every $y_0 \in F(x_0)$ there exists a sequence $\{y_n\}$ such that $y_n \in F(x_n)$ $\forall n$ and $y_n \to y_0$, $n \to \infty$.

If a set-valued map is sequentially lower semicontinuous at each point, it is called *sequentially lower semicontinuous*.

Exercise 17.3. *Prove that the sequential lower semicontinuity of a set-valued map F at a point x_0 is equivalent to the fact that*

$$\operatorname{dist}(y_0, F(x)) \to 0, \quad x \to x_0 \quad \forall y_0 \in F(x_0). \tag{17.1}$$

Consider some examples of sequentially upper and lower semicontinuous set-valued maps.

Exercise 17.4. *Let X be the interval $[0, 2]$, and Y the interval $[0, 1]$ with the natural metrics. Consider the set-valued maps F_1, F_2*

$$F_1(x) = \begin{cases} [0,1], & x \neq 1 \\ \{0\}, & x = 1 \end{cases}; \qquad F_2(x) = \begin{cases} \{0\}, & x \in [0,1), \\ [0,1], & x = 1, \\ \{1\}, & x \in (1,2]. \end{cases}$$

Prove that the set-valued map F_1 is sequentially lower semicontinuous but not sequentially upper semicontinuous at the point $x = 1$, whereas, conversely, the set-valued map

DOI 10.1515/9783110460308-018

F_2 is sequentially upper semicontinuous at $x = 1$ but not sequentially lower semicontinuous at the same point.

In addition to the above definitions, the following notions based on the deviation of sets and the Hausdorff metric are also used.

Definition 17.5. A set-valued map F is called h-upper semicontinuous at a point $x_0 \in X$, if for an arbitrary $\varepsilon > 0$ there exists such a $\delta > 0$, that

$$F(x) \subset O^Y(F(x_0), \varepsilon) \quad \forall x \in O^X(x_0, \delta). \tag{17.2}$$

Notice that inclusion (17.2) may be replaced with the inequality $h^+(F(x), F(x_0)) \leq \varepsilon$. If a set-valued map is h-upper semicontinuous at each point it is called *h-upper semicontinuous.*

Definition 17.6. A set-valued map F is called h-lower semicontinuous at a point $x_0 \in X$, if for an arbitrary $\varepsilon > 0$ there exists such a $\delta > 0$, that

$$O^Y(F(x), \varepsilon) \supset F(x_0) \quad \forall x \in O^X(x_0, \delta). \tag{17.3}$$

If a set-valued map is h-lower semicontinuous at each point it is called *h-lower semicontinuous.*

Consider an important example of a sequentially lower semicontinuous (h-lower semicontinuous) set-valued map.

Exercise 17.7. *Let F be a sequentially lower semicontinuous (h-lower semicontinuous) set-valued map. Take a point $x_0 \in X$ and let $A \subset F(x_0)$ be an arbitrary closed subset (e.g., $A = \{y_0\}$, where y_0 is a given point from $F(x_0)$). Set*

$$\tilde{F}(x) = \begin{cases} F(x), & x \neq x_0, \\ A, & x = x_0. \end{cases}$$

Prove that $\tilde{F}$ is also a sequentially lower semicontinuous (h-lower semicontinuous) set-valued map.

It is easy to verify that the h-upper semicontinuity of a set-valued map is equivalent to its sequential upper semicontinuity and so, in general, we need not distinguish these notions. But for the lower semicontinuous set-valued maps, in general, the implication is true only in one side. Namely, if a set-valued map is h-lower semicontinuous, then, as it is easy to see, it is sequentially lower semicontinuous. But, as the following example demonstrates, the converse is, generally speaking, not true.

Exercise 17.8. *Let X be the interval $[0, 1]$ and $Y = \mathbb{R}$ with the natural metrics. Consider the set-valued map*

$$F(x) = \begin{cases} [-1/x, 1/x], & x \in (0, 1] \\ \mathbb{R}, & x = 0. \end{cases}$$

Prove that the set-valued map F is sequentially lower semicontinuous but not h-lower semicontinuous at the point $x = 0$.

However, under an additional assumption of compactness, the sequential lower semicontinuity is equivalent to the h-lower semicontinuity. It is demonstrated by the following assertion.

Proposition 17.9. *Let a set-valued map F be sequentially lower semicontinuous at a point x_0 and the set $F(x_0)$ compact. Then the set-valued map F is h-lower semicontinuous at the same point.*

Proof. Suppose, to the contrary, that F is not h-lower semicontinuous at the point x_0. Then there exist $\varepsilon > 0$ and a sequence $\{x_n\}$ such that $x_n \to x_0$ and $F(x_0) \not\subset O^Y(F(x_n), \varepsilon)$ $\forall n$. So, there exist such $y_n \in F(x_0)$, that $\operatorname{dist}(y_n, F(x_n)) \geq \varepsilon$ $\forall n$. By virtue of compactness of the set $F(x_0)$, passing to a subsequence if necessary, we may assume, without loss of generality, that $y_n \to y_0$ for a certain point $y_0 \in F(x_0)$. Hence, by using the triangle inequality (15.9) we get $\operatorname{dist}(y_0, F(x_n)) \geq \varepsilon/2$ for all sufficiently large n. This contradicts (17.1), completing the proof. □

Therefore, for set-valued maps with compact values we also do not need to distinguish sequential lower semicontinuity and h-lower semicontinuity.

Definition 17.10. A set-valued map F is called continuous provided it is both h-upper and h-lower semicontinuous.

Let all the values $F(x)$ of a set-valued map F be closed bounded sets. Then from the above definition it follows that a set-valued map F is continuous if and only if it is continuous as the single-valued map between metric spaces X and $\mathcal{H}(Y)$ (i.e., it is continuous with respect to the Hausdorff metric).

Denote by $X \times Y$ the Cartesian product of metric spaces (X, ρ_X) and (Y, ρ_Y) with the metric defined by the relation

$$\rho((x_1, y_1), (x_2, y_2)) = \rho_X(x_1, x_2) + \rho_Y(y_1, y_2). \tag{17.4}$$

It is easy to see that the convergence in the metric space $X \times Y$ may be described as:

$$(x_n, y_n) \to (x_0, y_0) \Leftrightarrow x_n \to x_0,\ y_n \to y_0,\ n \to \infty.$$

Definition 17.11. The graph of a set-valued F is the set

$$\operatorname{gph} F = \{ (x, y) \in X \times Y : y \in F(x) \}.$$

Definition 17.12. A set-valued map F is called closed if its graph $\operatorname{gph} F$ is a closed set.

Theorem 17.13. *If a set-valued map F is sequentially upper semicontinuous, then it is closed.*

Proof. Suppose that a sequence $\{ (x_n, y_n) \}$ lies in gph F and converges to a point (x_0, y_0). Let us show that $(x_0, y_0) \in \operatorname{gph} F$. In fact, $y_n \in F(x_n)$ $\forall n$ and, as mentioned above, $x_n \to x_0$, $y_n \to y_0$, $n \to \infty$. By virtue of the sequential upper semicontinuity of the set-valued map F at the point x_0 we have $\operatorname{dist}(y_n, F(x_0)) \to 0$. By using inequality (15.9) and the closedness of the set $F(x_0)$, we get $y_0 \in F(x_0)$ and hence $(x_0, y_0) \in \operatorname{gph} F$. □

The following example shows that the assertion inverse to the statement of Theorem 17.13, generally speaking, is not true even if we suppose that all the values $F(x)$ are compact sets.

Example 17.14. Let both metric spaces X and Y be given as the set of all nonnegative real numbers with the natural metric. Consider the following set-valued map (which in fact is single-valued)

$$F(x) = \begin{cases} \{1/x\}, & x > 0 \\ \{0\}, & x = 0. \end{cases}$$

It is easy to see that the graph of this set-valued map is closed. At the same time, the set-valued map F is not sequentially upper semicontinuous at zero.

However, under the additional assumption concerning the compactness of the space Y, the closedness of a set-valued map implies its sequential upper semicontinuity.

Theorem 17.15. *Let the space Y be compact. Then a set-valued map F is sequentially upper semicontinuous if and only if it is closed.*

Proof. By virtue of Theorem 17.13 it is sufficient to show that if the set-valued map F is closed, then it is sequentially upper semicontinuous at each point $x_0 \in X$. Suppose that a sequence $\{x_n\}$ converges to a point x_0 and for a sequence $\{y_n\}$ we have $y_n \in F(x_n)$ $\forall n$. Let us demonstrate that $\operatorname{dist}(y_n, F(x_0)) \to 0$, $n \to \infty$.

In fact, suppose the contrary. Then there exists such an $\varepsilon > 0$, that, without loss of generality, after the transfer from the sequence $\{y_n\}$ to its subsequence, we have $\operatorname{dist}(y_n, F(x_0)) \geq \varepsilon$ $\forall n$. By virtue of compactness of the space Y, passing once more to a subsequence, we get $y_n \to y_0$ for a certain $y_0 \in Y$. Then $(x_n, y_n) \to (x_0, y_0)$ and $(x_n, y_n) \in \operatorname{gph} F$ $\forall n$. Hence $(x_0, y_0) \in \operatorname{gph} F$ by the closedness of the set gph F yielding

$$y_0 \in F(x_0) \Rightarrow \operatorname{dist}(y_n, F(x_0)) \leq \rho_Y(y_n, y_0),$$

resulting $\operatorname{dist}(y_n, F(x_0)) \to 0$, $n \to \infty$. The obtained contradiction completes the proof. □

Let us present the generalization of Theorem 17.15.

Definition 17.16. A set-valued map F is called locally compact if for each point $x \in X$ there exists such a neighborhood $O(x)$, that the set $F(O(x))$ is relatively compact (i.e., it is contained in a certain compact set).

Theorem 17.17. *Let a set-valued map F be closed and locally compact. Then it is sequentially upper semicontinuous.*

The proof of this assertion follows the same lines as the proof of Theorem 17.15 (for details see [17], Theorem 1.2.32).

Let us return to Exercise 17.4. There we presented the set-valued maps F_1 and F_2, the first of which was sequentially lower semicontinuous but not sequentially upper semicontinuous at a certain (unique) point, and, conversely, the set-valued map F_2 which was sequentially upper semicontinuous but lost the lower semicontinuity at a certain (also unique) point. It turns out that this "extensiveness", in a certain sense, of the set of continuity points is not accidental. Namely, the following deep assertion, proved in [43] (see also [44], § 43) holds true.

Theorem 17.18. *Suppose that a metric space X is complete, a set-valued map F is compact-valued and is either sequentially upper or sequentially lower semicontinuous. Then there exists a subset $A \subset X$ which may be presented as the intersection of a denumerable number of open everywhere dense subsets and such that F is continuous on A.*

Notice that by the Baire[1] theorem (see [30]) the mentioned set A is everywhere dense in X.

Let us also present the following interesting assertion obtained in [45].

Theorem 17.19. *Suppose that $Y = \mathbb{R}^n$, set-valued maps F and G have compact convex values, F is sequentially lower semicontinuous, G is sequentially upper semicontinuous and $G(x) \subset F(x)$ for all $x \in X$. Then there exists such a continuous set-valued map H with compact convex values that $G(x) \subset H(x) \subset F(x)$ for all $x \in X$.*

Now, consider a collection of set-valued maps F_j, $j \in J$, where J is a certain set of indices. Define new set-valued maps, the union and the intersection of the initial set-valued maps by the relations

$$\left(\bigcup_{j\in J} F_j\right)(x) = \bigcup_{j\in J} F_j(x), \quad \left(\bigcap_{j\in J} F_j\right)(x) = \bigcap_{j\in J} F_j(x).$$

In so doing, we assume that in the union formula the set of indices J is finite and the intersection is well defined, i.e., $\cap_{j\in J} F_j(x) \neq \emptyset \ \forall x$.

Exercise 17.20. *Prove that the following relations hold true*

$$\operatorname{gph}\left(\bigcup_{j\in J} F_j\right) = \bigcup_{j\in J} \operatorname{gph} F_j, \quad \operatorname{gph}\left(\bigcap_{j\in J} F_j\right) = \bigcap_{j\in J} \operatorname{gph} F_j.$$

1 René-Louis Baire (1874–1932), a French mathematician.

Lemma 17.21. *Set $F = \bigcup_{j \in J} F_j$.*

(1) *Let all the set-valued maps F_j be sequentially upper semicontinuous. Then F is also sequentially upper semicontinuous.*

(2) *Let all the set-valued maps F_j be sequentially lower semicontinuous (h-lower semicontinuous). Then F is also sequentially lower semicontinuous (respectively, h-lower semicontinuous).*

(3) *Let all the set-valued maps F_j be closed. Then F is also closed.*

Proof. Let us start with assertion (1). Suppose, to the contrary, that the set-valued map F is not sequentially upper semicontinuous at a point $x_0 \in X$. Then there exist such $\varepsilon > 0$ and sequences $\{x_n\}$, $\{y_n\}$ such that

$$x_n \to x_0, \; y_n \in F(x_n) \; \forall n, \quad \operatorname{dist}(y_n, F(x_0)) \geq \varepsilon \; \forall n.$$

Passing to a subsequence, attain that for a certain number j we will have $y_n \in F_j(x_n) \; \forall n$. But $\operatorname{dist}(y_n, F_j(x_0)) \geq \operatorname{dist}(y_n, F(x_0)) \geq \varepsilon \; \forall n$, contrary to the sequential upper semicontinuity of the set-valued map F_j at the point x_0.

The validity of assertion (2) follows from the definitions and we get assumption (3) from Exercise 17.20. □

The proved lemma shows that the union (of a finite number) of sequentially upper semicontinuous [sequentially lower semicontinuous, h-lower semicontinuous] set-valued maps is sequentially upper semicontinuous [respectively, sequentially lower semicontinuous, h-lower semicontinuous]. The situation with the intersection of set-valued maps is much more complicated: in general, this operation preserves neither sequential upper semicontinuity nor sequential lower semicontinuity even if the set of indices is finite (although it preserves the closedness property that follows immediately from Exercise 17.20). Let us illustrate this by examples.

Example 17.22 (see [17], Example 1.3.6). Let X be the interval $[0, \pi]$ and $Y = \mathbb{R}^2$ with the natural metric. The set-valued map F_1 is constant and for all x the set $F_1(x)$ is the upper half of the unit disc lying in the first and the second orthants and $F_2(x)$ is the diameter of the unit disc forming the angle x with the real axis, i.e.,

$$F_2(x) = \{ (y_1, y_2) \in \mathbb{R}^2 : y_1 = r\cos x, \; y_2 = r\sin x, \; r \in [-1, 1] \}.$$

The set-valued maps F_1, F_2 are obviously sequentially lower semicontinuous (they are even continuous) but their intersection $\bigcap_{j=1}^{2} F_j$ is not sequentially lower semicontinuous at the points 0 and π. (Prove it as an exercise).

In the following example the intersection of two sequentially upper semicontinuous set-valued maps is not sequentially upper semicontinuous.

Example 17.23. Let X be the interval $[0, 1]$, and $Y = \mathbb{R}_+$ a nonnegative real axis with the natural metric. Define the set-valued maps F_1, F_2 by the relations

$$F_1(x) = F_2(x) = \begin{cases} \{n\}, & x \in \left(\frac{1}{n+1}, \frac{1}{n}\right), \quad n = 1, 2, \dots \\ \{n-1, n\}, & x = \frac{1}{n}, \quad n = 2, 3, \dots \\ \{1\}, & x = 1, \end{cases}$$

and

$$F_1(0) = \{0, 1, 2, \dots, n, \dots\}, \quad F_2(0) = \left\{0, 1 + \frac{1}{2}, 2 + \frac{1}{3}, \dots, n + \frac{1}{n+1}, \dots\right\}.$$

It is obvious that both set-valued maps F_1, F_2 are sequentially upper semicontinuous. At the same time, for their intersection $F = \bigcap_{j=1}^{2} F_j$ we get $F(0) = \{0\}$, $F(x) = F_1(x)\ \forall x \in (0, 1]$ and hence, as it is easy to see (prove it!) the set-valued map F is not sequentially upper semicontinuous.

Nevertheless, for the intersection of set-valued maps we have the following simple, but important assertion (see Theorem 1.3.3 from [17]).

Theorem 17.24. *Let two set-valued maps F_1 and F_2 be given; the set-valued map F_1 is closed and the set-valued map F_2 is sequentially upper semicontinuous and compact-valued. Then if their intersection $F = F_1 \bigcap F_2$ is well defined then it is sequentially upper semicontinuous.*

Proof. Take an arbitrary point $x_0 \in X$ and prove that the set-valued map F is sequentially upper semicontinuous at x_0. The proof is similar to those of Lemma 17.15. In fact, assume that a sequence $\{x_n\}$ converges to a point x_0 and $y_n \in F(x_n)\ \forall n$. Let us prove that $\operatorname{dist}(y_n, F(x_0)) \to 0$, $n \to \infty$. Suppose the contrary. Then there exists such an $\varepsilon > 0$, that, after passing to a subsequence, we get $\operatorname{dist}(y_n, F(x_0)) \geq \varepsilon\ \forall n$.

But $y_n \in F_2(x_n)\ \forall n$ and hence $\operatorname{dist}(y_n, F_2(x_0)) \to 0$ by virtue of the sequential upper semicontinuity of F_2. Let us prove the existence of such $y_0 \in F_2(x_0)$, that after passing to a subsequence, we have $y_n \to y_0$. In fact, from the proved it follows that for each n there exists such a $\tilde{y}_n \in F_2(x_0)$, that $\rho_Y(\tilde{y}_n, y_n) \to 0$. By virtue of the compactness of the set $F_2(x_0)$, passing to the subsequence, we get $\tilde{y}_n \to y_0$ for a certain $y_0 \in F_2(x_0)$. From here, by the triangle inequality we have $y_n \to y_0$, proving the desired.

Moreover, $y_n \in F_1(x_n)\ \forall n$, implying by the closedness of F_1 that $y_0 \in F_1(x_0)$. So, it is proved that $y_0 \in F(x_0)$ and hence for a given subsequence we have $\operatorname{dist}(y_n, F(x_0)) \to 0$, contrary to the choice of $\varepsilon > 0$. □

The proved theorem implies the following assertion.

Corollary 17.25. *Let set-valued maps F_j, $j \in J$ be closed, their intersection $F = \bigcap_{j \in J} F_j$ is well defined and, at least one of them, $F_{\bar{j}}$ be sequentially upper semicontinuous and compact valued. Then the intersection F is sequentially upper semicontinuous.*

Proof. In fact, by virtue of Proposition 17.20 the set-valued map F is closed. It is evident that $F = F \bigcap F_{\bar{j}}$, where $F_{\bar{j}}$ is compact valued. It remains to apply Theorem 17.24 to set-valued maps $F_1 = F$ and $F_2 = F_{\bar{j}}$. □

Sequentially upper semicontinuous set-valued maps frequently arise in applications. Consider two examples.

Consider a function $f: X \times Y \to \mathbb{R}$, and a set-valued map Φ on X with compact values $\Phi(x) \subset Y$. For each fixed x, the function $f(x, \cdot)$ achieves the maximum on the compact $\Phi(x)$. Denote this maximum by $m(x)$ and set

$$M(x) = \{ y \in \Phi(x) : f(x, y) = m(x) \}.$$

The necessity of the study of the function m and the set-valued map M, which are sometimes called *marginal*, very often arises in the theory of games, mathematical economics and other applications (see, e.g., [17]). Their main properties are expressed by the following assertion which is called *the principle of continuity of optimal solutions.*

Proposition 17.26. *Suppose that the function f and the set-valued map Φ are continuous. Then the function m is continuous, the sets $M(x)$ are compact for all x, and the set-valued map M is sequentially upper semicontinuous.*

Proof. Take an arbitrary point $x_0 \in X$. Let a sequence $\{x_n\}$ converge to the point x_0 and $y_n \in M(x_n)$ $\forall n$. Then $y_n \in \Phi(x_n)$ $\forall n$. Hence, by virtue of the sequential upper semicontinuity of the set-valued map Φ we have $\operatorname{dist}(y_n, \Phi(x_0)) \to 0$. Therefore, for each number n there exists $\tilde{y}_n \in \Phi(x_0)$ such that $\rho_Y(y_n, \tilde{y}_n) \to 0$.

Let us prove that $\operatorname{dist}(y_n, M(x_0)) \to 0$. Indeed, suppose the contrary. Then there exists such an $\varepsilon > 0$, that, after passing to a subsequence we have $\operatorname{dist}(y_n, M(x_0)) > \varepsilon$ $\forall n$. By the compactness of the set $\Phi(x_0)$, passing once more to a subsequence, we will suppose that $\tilde{y}_n \to y_0$ for a certain $y_0 \in \Phi(x_0)$. But then also $y_n \to y_0$.

Let us show that $y_0 \in M(x_0)$. Indeed, take an arbitrary $y \in \Phi(x_0)$. By the sequential lower semicontinuity of the set-valued map Φ there exists such a sequence sequence $\{\bar{y}_n\}$ convergent to y that $\bar{y}_n \in \Phi(x_n)$ $\forall n$. Then $f(x_n, \bar{y}_n) \le f(x_n, y_n)$ $\forall n$. Passing in this inequality to the limit as $n \to \infty$, we get $f(x_0, y) \le f(x_0, y_0)$. From here, by the arbitrariness of the point $y \in \Phi(x_0)$ we conclude that $y_0 \in M(x_0)$. The obtained contradiction proves that $\operatorname{dist}(y_n, M(x_0)) \to 0$. So, we justified the sequential upper semicontinuity of M at the point x_0, as well as the closedness of the set $M(x_0)$.

Let us prove the continuity of the function m at the point x_0. By virtue of that proved above, for the constructed subsequences we have $m(x_n) = f(x_n, y_n)$, $m(x_0) = f(x_0, y_0)$, that yields $m(x_n) \to m(x_0)$. But by virtue of the above reasoning, from each subsequence of the initial sequence $\{x_n\}$ we can choose a subsequence $\{x_{n_k}\}$, for which $m(x_{n_k}) \to m(x_0)$, $k \to \infty$, that proves the continuity of m at the point x_0. □

The set-valued maps $\Phi = F_1$ and $\Phi = F_2$, where F_1 and F_2 are taken from Exercise 17.4, and the function $f(x, y) \equiv y$ demonstrate that in the proved proposition we cannot weaken the assumptions concerning the set-valued map Φ, replacing its continuity with the upper or lower sequential semicontinuity. In both cases the corresponding set-valued map M is not upper (and lower) sequentially semicontinuous at the point $x_0 = 1$, and the function m is discontinuous at this point (verify it as an exercise).

We will take the second example from convex analysis (see Part I). Let a convex proper function $f: \mathbb{R}^n \to \bar{\mathbb{R}}$ be bounded above in a certain neighborhood $O(x_0)$ of a given point x_0. Its subdifferential may be considered as the set-valued map which assigns to each point $x \in O(x_0)$ the subdifferential $\partial f(x)$ of this function at the point x.

Let us show that the subdifferential set-valued map ∂f is sequentially upper semicontinuous at x_0. To this end, let us prove first that it is closed on $O(x_0)$. Let a sequence $\{x_n\}$ converge to a point $x \in O(x_0)$, and a sequence $\{y_n^*\}$ such that $y_n^* \in \partial f(x_n)$ $\forall n$ converge to a point y^*. Then

$$f(\xi) \geq f(x_n) + \langle y_n^*, \xi - x_n \rangle \quad \forall \xi.$$

But from Theorem 6.8 we know that the function f is continuous on $O(x_0)$. So, passing in the above inequality for each fixed ξ to the limit as $n \to \infty$, we get $y^* \in \partial f(x)$ and hence the subdifferential set-valued map ∂f is closed on $O(x_0)$.

Further, repeating the reasoning from the proof of Theorem 9.9, we can find such a neighborhood $O'(x_0) \subseteq O(x_0)$ of the point x_0 that all the sets $\partial f(x)$, $x \in O'(x_0)$ are contained in a certain ball. So, the set-valued map ∂f is compact on $O'(x_0)$ and, by Theorem 17.17 it is sequentially upper semicontinuous at x_0.

Notice that this continuity property of the subdifferential plays an important role in applications.

Now, let us return to the notions of semicontinuity of set-valued maps. Apart from the above introduced definitions of upper and lower sequential and h-semicontinuity the following approaches, expressed in topological terms, are commonly used.

Definition 17.27. A set-valued map F is called upper semicontinuous at a point $x_0 \in X$, if for each neighborhood O^Y of the set $F(x_0)$ there exists such a neighborhood O^X of the point x_0, that $F(O^X) \subset O^Y$.

Definition 17.28. A set-valued map F is called lower semicontinuous at a point $x_0 \in X$, if for each open set $O^Y \subset Y$ such that $O^Y \bigcap F(x_0) \neq \emptyset$ there exists a neighborhood O^X of the point x_0 such that $F(x) \bigcap O^Y \neq \emptyset$ for every $x \in O^X$.

These definitions are useful in situations when X and Y are only topological (not metric) spaces. In cases when X and Y are metric spaces, the lower semicontinuity is equivalent to the sequential lower semicontinuity (see [17], Theorem 1.2.20).

For upper semicontinuous set-valued maps the case is somewhat more complicated. From the upper semicontinuity of a set-valued map in the sense of Definition 17.27 follows its sequential upper semicontinuity but the inverse implication is valid only under the additional assumption of the compactness of the value $F(x_0)$ (see [17], Theorem 1.2.20). As the following example demonstrates, this condition is essential.

Exercise 17.29. *Consider the set-valued map F given on the plane* $\mathbb{R}^2$ *by the formula* $F(x, y) = (x, y) + \mathbb{R}^2_+$, *where* $\mathbb{R}^2_+ = \{(a, b) : a \geq 0,\ b \geq 0\}$. *Prove that F is sequentially upper semicontinuous, but it is not upper semicontinuous.*

Summing up, we can say that for compact-valued maps of metric spaces the notions of sequential upper semicontinuity, h-upper semicontinuity and upper semicontinuity (as well as sequential lower semicontinuity, h-lower semicontinuity and lower semicontinuity) are equivalent.

Let us also mention the following useful property of upper semicontinuous compact-valued maps generalizing the well-known topological property of continuous maps.

Proposition 17.30. *Let F be an upper semicontinuous compact-valued map from a metric space X into a metric space Y. If a set* $A \subset X$ *is compact, then its image* $F(A) = \bigcup_{a \in A} F(a)$ *is also compact.*

Proof. Let $\{y_n\} \subset F(A)$ be an arbitrary sequence. Take a sequence $\{x_n\} \subset A$ such that $y_n \in F(x_n)\ \forall n$. Since A is compact we can assume, without loss of generality, that $x_n \to x_0 \in A$. But then the sequential upper semicontinuity of F implies $\operatorname{dist}(y_n, F(x_0)) \to 0$. From the compactness of the set $F(x_0)$ it easily follows that there exists a subsequence $\{y_{n_k}\}$ tending to $y_0 \in F(x_0) \subset F(A)$ that concludes the proof. □

Notice that the topological notion of upper semicontinuity (Definition 17.27) has its peculiarities. For example, if given set-valued maps F_j, $j = \overline{1, m}$ are upper semicontinuous in the sense of Definition 17.27 and their intersection is well defined then they are also upper semicontinuous in the same sense (see [17], Theorem 1.3.2). At the same time, Example 17.23 demonstrates that for sequentially upper semicontinuous set-valued maps this is not so.

In concluding this section, let us make important terminological remarks. There are classical notions of upper and lower semicontinuous functions in the theory of functions of a real variable (see Definition 6.1 in Part I). The connection of these notions with our definitions of upper and lower semicontinuous set-valued maps may be described in the following way. Suppose that the space Y is separable and let $\{r_n\}$ be any its denumerable everywhere dense subsets. For a set-valued map F define the functions $\varphi_n : X \to \mathbb{R}_+$, $\varphi_n(x) = \operatorname{dist}(r_n, F(x)$. Then the lower semicontinuity of F is equivalent to the upper semicontinuity of all functions φ_n. Under the additional assumption of compactness of the space Y we have the dual assertion: the upper semicontinuity of F is equivalent to the lower semicontinuity of all functions φ_n (see [17], Theorems 1.2.45 and 1.2.47).

Notice also that if we are considering a single-valued map as the set-valued one, then its upper as well as lower semicontinuity is obviously equivalent to classical continuity.

18 A base of topology of the space $\mathcal{H}_c(X)$

Let us make a small topological deviation which we will need in the following. As previously defined, the collection $\mathcal{H}_c(X)$ consisting of all nonempty compact subsets of a metric space X endowed with the Hausdorff metric h is a metric space We would like to describe a base of the topology on $\mathcal{H}_c(X)$ generated by this metric.

For an arbitrary open set $G \subset X$ consider the collections of subsets

$$\mathcal{F}(G) = \{ K \in \mathcal{H}_c(X) : K \bigcap G \neq \emptyset \},$$
$$\mathcal{L}(G) = \{ K \in \mathcal{H}_c(X) : K \subset G \}.$$

In the space $\mathcal{H}_c(X)$ consider the topology which is the weakest among all those containing collections $\mathcal{F}(G)$ and $\mathcal{L}(G)$ for all open $G \subset X$. This topology is called the Vietoris[1] topology. We will show that the Vietoris topology is generated by the Hausdorff metric and hence sets of type $\mathcal{F}(G)$ and $\mathcal{L}(G)$ form a prebase of the topology of the metric space $\mathcal{H}_c(X)$.

Proposition 18.1. *Let a space $G \subset X$ be open in X. Then the sets $\mathcal{F}(G)$ and $\mathcal{L}(G)$ are open in $\mathcal{H}_c(X)$.*

Proof. Take an arbitrary $K \in \mathcal{F}(G)$. Then there exists a point $x \in K \cap G$. Choose $\varepsilon > 0$ such that $O(x, \varepsilon) \subset G$. Take an arbitrary $\tilde{K} \in \mathcal{H}_c(X)$ such that $h(K, \tilde{K}) < \varepsilon$. Then $K \subset O(\tilde{K}, \varepsilon)$. So, there exists a point $\tilde{x} \in \tilde{K}$ such that $\rho_X(x, \tilde{x}) < \varepsilon$. Therefore, $\tilde{K} \bigcap G \neq \emptyset$ and hence $\tilde{K} \in \mathcal{F}(G)$. We see that K is an internal point of the family $\mathcal{F}(G)$ and hence it is open in $\mathcal{H}_c(X)$.

Take now an arbitrary $K \in \mathcal{L}(G)$. Denote by C the complement to the open set G. Then the set C is closed and does not intersect with the compact set K. Hence there exists such an $\varepsilon > 0$ that $O(K, \varepsilon) \bigcap C = \emptyset$ [2] and hence $O(K, \varepsilon) \subset G$. Take an arbitrary $\tilde{K} \in \mathcal{H}_c(X)$ such that $h(K, \tilde{K}) < \varepsilon$. Then $\tilde{K} \subset O(K, \varepsilon) \subset G$. This means that K is an internal point of the family $\mathcal{F}(G)$ and hence it is open in $\mathcal{H}_c(X)$. □

Theorem 18.2. *Sets of type $\mathcal{F}(G)$ and $\mathcal{L}(G)$, where G are all possible open subsets of X, form a prebase of the topology of the metric space $\mathcal{H}_c(X)$.*

Proof. Notice that to demonstrate that a system of open sets is a base of a topology, it is sufficient to show that for each point x and each of its neighborhoods $O(x)$ there exists an element of this system containing the point x and contained itself in $O(x)$ (see, e.g., [27]). From here it follows, taking into account Proposition 18.1, that it is sufficient to

1 Leopold Vietoris (1891–2002), an Austrian mathematician.

2 If a nonempty compact subset of a metric space does not intersect a closed one, then a certain ε-neighborhood of this compact set also does not intersect this closed set. This assertion, which is obviously equivalent to the fact that each open neighborhood of a compact set contains its certain ε-neighborhood, can be easily proved as an exercise.

DOI 10.1515/9783110460308-019

construct, for arbitrary $K \in \mathcal{H}_c(X)$ and $\varepsilon > 0$, such a subset $\mathcal{N} \subset \mathcal{H}_c(X)$, that $\mathcal{N}$ can be represented in the form of the intersection of a finite number of sets of type $\mathcal{F}(G_j)$ or $\mathcal{L}(G_i)$, $K \in \mathcal{N}$, and $\mathcal{N}$ is contained in a closed ε-neighborhood of the point K.

Let us do it. In fact, the compact set K is totally bounded and hence it has a finite $\varepsilon/2$-net $x_1, \dots, x_n$. Set

$$G_i = O(x_i, \varepsilon/2), \quad \mathcal{N} = \mathcal{L}(O(K, \varepsilon)) \bigcap \mathcal{F}(G_1) \bigcap \cdots \bigcap \mathcal{F}(G_n),$$

and prove that the set $\mathcal{H} \subset \mathcal{H}_c(X)$ is the desirable one.

Indeed, $K \in \mathcal{N}$. Take an arbitrary point $\tilde{K} \in \mathcal{N}$. We have $\tilde{K} \subset O(K, \varepsilon)$. In addition, $\tilde{K} \bigcap G_i \neq \emptyset$, $i = \overline{1, n}$. Hence there exist points $g_i \in \tilde{K} \bigcap G_i$. Taking into account that, by construction $\bigcup_{i=1}^{n} O(x_i, \varepsilon/2) \supset K$, we get

$$O(\tilde{K}, \varepsilon) \supset \bigcup_{i=1}^{n} O(g_i, \varepsilon) \supset \bigcup_{i=1}^{n} O(x_i, \varepsilon/2) \supset K.$$

Therefore, $\tilde{K} \subset O(K, \varepsilon)$, $K \subset O(\tilde{K}, \varepsilon)$ yielding $h(K, \tilde{K}) \leq \varepsilon$. The constructed set $\mathcal{N}$ is desired. □

19 Measurable set-valued maps. Measurable selections and measurable choice theorems

In this section we will assume that a metric space X is endowed with a denumerably additive regular and nonnegative measure μ, such that $\mu(X)$ is finite.[1] Recall that each open or closed set is measurable and the regularity of a nonnegative measure μ means that for every measurable set A and each $\varepsilon > 0$ there exist such a closed set $C \subset A$ and such an open set $O \supset A$, that $\mu(O \setminus C) < \varepsilon$. An empty set is measurable.

In this section we will assume that the metric space Y is separable and, moreover, all considered set-valued maps are compact valued. As already mentioned, for these maps all definitions of upper (lower) semicontinuity are equivalent.

Definition 19.1. A set-valued map $F\colon X \to \mathcal{H}_c(Y)$ is called measurable if for each open set $O \subset \mathcal{H}_c(Y)$ its pre-image $F^{-1}(O) = \{x \in X\colon F(x) \in O\}$ is measurable.

Let us present a criterion of measurability of a set-valued map $F\colon X \to \mathcal{H}_c(Y)$. To do so, define "the complete pre-image" of a subset $A \subset Y$ by the relation

$$F^{-}(A) = \{x \in X\colon F(x) \bigcap A \neq \emptyset\}.$$

Theorem 19.2. *The following statements are equivalent:*

(1) *A set-valued map F is measurable;*
(2) *The set $F^{-}(A)$ is measurable for each measurable set $A \subset Y$;*
(3) *The set $F^{-}(A)$ is measurable for each open set $A \subset Y$.*

Proof. (a) Let the map F be measurable. Let us prove item (3). By virtue of Theorem 15.3 the space $\mathcal{H}_c(Y)$ is separable. So, by the Luzin[2] theorem (see [24], Theorem 1.4.19) for an arbitrary $\varepsilon > 0$ there exists such a closed subset $X_\varepsilon \subset X$, that $\mu(X \setminus X_\varepsilon) < \varepsilon$, and the restriction of the map F to the set X_ε is continuous.

Let a set $A \subset Y$ be open. If the set $X_\varepsilon \bigcap F^{-}(A)$ is nonempty, then take an arbitrary point $\bar{x} \in X_\varepsilon \cap F^{-}(A)$ and choose a point $\bar{y} \in F(\bar{x}) \bigcap A$. Since A is open, there exists such a $\delta > 0$, that $O^Y(\bar{y}, \delta) \subset A$. Since the set-valued map F is continuous on X_ε, there exists such an $\eta > 0$, that for all $x \in X_\varepsilon$ with $\rho_X(x, \bar{x}) < \eta$ we have $F(\bar{x}) \subset O^Y(F(x), \delta)$ and hence $\bar{y} \in O^Y(F(x), \delta)$. Therefore, for each such x there exists a point $y \in F(x)$ with $y \in O^Y(\bar{y}, \delta) \subset A$ and so $y \in F(x) \bigcap A$, implying $x \in F^{-}(A) \bigcap X_\varepsilon$. Thus, the intersection of an η-neighborhood of a point $\bar{x}$ with X_ε belongs to $F^{-}(A)$. This means that if the set

1 To read this section, it is helpful to be preliminary familiar with the main concepts of measure theory (see, e.g., [27], Ch. V, or [24], Ch. 1). If the reader does not want to go into the details of measure theory, it is sufficient to assume that X is the interval $[0, 1]$ with the natural metric and the Lebesgue measure. The presentation in this section follows [24], Sect. 1.7. Notice, however, that in that monograph it is supposed that the spaces X and Y are compact, which simplifies the reasoning.
2 Nikolai Nikolaevich Luzin (1883–1950), a Russian mathematician.

DOI 10.1515/9783110460308-020

$F^{-}(A) \bigcap X_{\varepsilon}$ is nonempty, then it can be represented as the intersection of an open set with a closed one, and hence it is measurable. Since $\mu(X \setminus \bigcup_{j=1}^{\infty} X_{1/j}) = 0$, by virtue that proved above, the set $F^{-}(A)$ can be represented as the intersection of a denumerable number of measurable sets and hence it is measurable.

(b) Let us prove that item (3) implies item (2). Let $A \subset Y$ be a closed set. Then $A = \bigcap_{j=1}^{\infty} O^{Y}(A, 1/j)$ (prove this as an exercise!) Let us show that

$$F^{-}(A) = \bigcap_{j=1}^{\infty} F^{-}(O^{Y}(A, 1/j)). \tag{19.1}$$

Indeed, by virtue of evident properties of "the complete pre-image" we have

$$F^{-}(A) = F^{-}\left(\bigcap_{j=1}^{\infty} O^{Y}(A, 1/j)\right) \subset \bigcap_{j=1}^{\infty} F^{-}(O^{Y}(A, 1/j)).$$

Let us prove the inverse inclusion

$$\bigcap_{j=1}^{\infty} F^{-}(O^{Y}(A, 1/j)) \subset F^{-}(A). \tag{19.2}$$

In fact, let $x \in F^{-}(O^{Y}(A, 1/j))$ for all j. Then

$$F(x) \bigcap O^{Y}(A, 1/j) \neq \emptyset \ \forall j \quad \Rightarrow \forall j \ \exists y_j : y_j \in F(x), \ y_j \in O^{Y}(A, 1/j).$$

By virtue of compactness of the set $F(x)$, passing to a subsequence if necessary, we will assume that $y_j \to y_0 \in F(x)$. Since A is closed, it is evident that $y_0 \in A$ and hence $x \in F^{-}(A)$, proving (19.2), together with (19.1).

From item (3) it follows that each of the sets $F^{-}(O^{Y}(A, 1/j))$ is measurable and hence $F^{-}(A)$ is also measurable as the intersection of a denumerable number of measurable sets.

(c) Now, demonstrate that item (3) implies item (1). Take an arbitrary open set $O \subset \mathcal{H}_c(Y)$. Since, by Theorem 15.3 the space $\mathcal{H}_c(Y)$ is separable, its open subset O is also a separable metric space. Therefore, it possesses a denumerable base of the topology and hence from each open covering of O it is possible to select a denumerable subcovering (see, e.g., [27]).

Let $\mathcal{B}$ be an arbitrary base of the topology of the space $\mathcal{H}_c(Y)$. Then for each $x \in O$ there exists such a set $O(x) \in \mathcal{B}$, that $x \in O(x) \subset O$ (see, e.g., [27]). Selecting a denumerable subcovering from an open covering $\bigcup_{x \in O} O(x)$, we get such a sequence $\{O_i\}$ of open sets $O_i \in \mathcal{B}$ that $O = \bigcup_{i=1}^{\infty} O_i$.

But, by Theorem 18.1, as a base $\mathcal{B}$ we can take all possible finite intersections of sets of type $\mathcal{F}(G)$ or type $\mathcal{L}(G)$, where G runs over all possible open sets of Y. Thus, the set O can be represented in the form of a denumerable union of sets, each of which is the intersection of a finite number of sets being the sets of type $\mathcal{F}(G)$ or type $\mathcal{L}(G)$, where G is an open set of Y. Therefore it is sufficient to prove the measurability of each of sets $F^{-1}(\mathcal{F}(G))$ and $F^{-1}(\mathcal{L}(G))$. Let us do it.

We have

$$F^{-1}(\mathcal{F}(G)) = \{x \in X : F(x) \in \mathcal{F}(G)\}$$
$$= \Big\{x \in X : F(x) \bigcap G \neq \emptyset\Big\} = F^{-}(G),$$
$$F^{-1}(\mathcal{L}(G)) = \{x \in X : F(x) \subset G\}$$
$$= X \setminus \Big\{x \in X : F(x) \bigcap (Y \setminus G) \neq \emptyset\Big\} = X \setminus F^{-}(Y \setminus G).$$

The set $F^{-}(G)$ is measurable by virtue of item (3). The set $F^{-}(Y \setminus G)$ is measurable since $(Y \setminus G)$ is closed and, by that proved above, item (3) implies item (2). The measurability of required sets is proved.

(d) Suppose that item (2) holds true and prove item (3). Let $A \subset Y$ be an open set. It is clear that it is separable. So, it has a denumerable base of topology and hence, it is possible to select a denumerable subcovering from each open covering of A (see, e.g., [27]). This immediately implies the existence of such $a_n \in A$ and $\varepsilon_n > 0$, $n = 1, 2, \ldots,$ that

$$A \subset \bigcup_{n=1}^{\infty} O^{Y}(a_n, \varepsilon_n) \subset \bigcup_{n=1}^{\infty} B^{Y}(a_n, \varepsilon_n) \subset A.$$

Therefore

$$F^{-}(A) = \bigcup_{n=1}^{\infty} F^{-}(B^{Y}(a_n, \varepsilon_n))$$

and hence $F^{-}(A)$ is measurable as the union of a denumerable number of measurable sets. □

Lemma 19.3. *If a set-valued map $F: X \to \mathcal{H}_c(Y)$ is upper semicontinuous, then it is measurable.*

Proof. Take an arbitrary closed set $A \subset Y$ and show that the set $F^{-}(A)$ is closed. In fact, it is easy to see that $F^{-}(A)$ is the complement in X to the set $\{x \in X : F(x) \subset CA\}$, where CA denotes the complement to A in Y. But, according to Definition 17.27, this last set is open, proving the desired. The measurability of the set-valued map F follows from Theorem 19.2. □

Lemma 19.4. *If a set-valued map $F: X \to \mathcal{H}_c(Y)$ is lower semicontinuous, then it is measurable.*

Proof. Let A be an arbitrary open subset of Y. The openness of the set $F^{-}(A)$ follows immediately from Definition 17.28 and then Theorem 19.2 yields the measurability of the set-valued map F. □

Lemma 19.5. *Let set-valued maps*

$$F_i : X \to \mathcal{H}_c(Y), \; i = 1, 2, \ldots,$$

such that

$$F(x) = \bigcap_{i=1}^{\infty} F_i(x) \neq \emptyset \ \forall x.$$

be measurable. Then the intersection $F\colon X \to \mathcal{H}_c(Y)$ *is also measurable.*

Proof. Take an arbitrary $\varepsilon > 0$. It is easy to verify (do it as an exercise!) that from the Lusin theorem it follows that there exists such a closed subset $X_\varepsilon \subset X$, that $\mu(X \setminus X_\varepsilon) < \varepsilon$ and the restriction of each set-valued map F_i to X_ε is continuous. Then, by virtue of the corollary to Theorem 17.24, the restriction of the map F to the set X_ε is upper semicontinuous and hence, by Lemma 19.3 it is measurable. From here, by virtue of the arbitrariness of $\varepsilon > 0$ it follows that the map F is measurable on the whole space X. □

Let a set-valued map F be given. A single-valued map f such that $f(x) \in F(x)$ for all $x \in X$ is called *the selection of* F. This notion raises a natural question: does a selection possessing any specific properties exist? For example, measurability, continuity, Lipschitz property, etc. More precisely, can certain properties of a set-valued map be inherited by any of its selections? This question is interesting both from the theoretical side as well as from the point of view of applications. Here we will study this question for measurable set-valued maps and in the following we will consider it for other classes of maps.

The following selection theorem is due to K. Kuratowski[3] and C. Ryll-Nardzewski[4] [46].

Theorem 19.6. *If a set-valued map F is measurable then it admits a measurable selection.*

Proof. Take an everywhere dense subset $\{y_1, y_2, \dots\}$ in Y. For each $x \in X$ set

$$F_0(x) = F(x)$$

and

$$F_{n+1}(x) = \{y \in F_n(x)\colon \rho_Y(y, y_{n+1}) = \operatorname{dist}(F_n(x), y_{n+1})\}, \quad n = 0, 1, \dots$$

Each of the sets $F_n(x)$ is nonempty and compact. It follows from the fact that if $K \subset Y$ is a compact set, then, by the Weierstrass theorem, for each fixed $\bar{y} \in Y$ the continuous function $\phi(y) = \rho_Y(y, \bar{y})$ achieves its minimum on the set K and the set of minimizing points is closed.

Fix an arbitrary $\varepsilon > 0$. By the Luzin theorem, there exists such a closed subset $X_\varepsilon^0 \subset X$, that $\mu(X \setminus X_\varepsilon^0) < \varepsilon/2$, and the restriction of the map F_0 to X_ε^0 is continuous.

3 Kazimierz Kuratowski (1896–1980), a Polish mathematician.
4 Czeslaw Ryll-Nardzewski (1926–2015), a Polish mathematician.

Construct by induction such closed sets X_ε^i, that for all integers $i \geq 1$ we have: $X_\varepsilon^i \subset X_\varepsilon^{i-1}$, $\mu(X_\varepsilon^{i-1} \setminus X_\varepsilon^i) \leq 2^{-(i+1)}\varepsilon$ and the restriction of the set-valued map F_i to the set X_ε^i is continuous.

Indeed, the set X_ε^0 is constructed already. Suppose that the sets X_ε^j are also constructed for all $j \leq i$. Let us construct the set X_ε^{i+1}. Consider the set X_ε^i. By Lemma 17.26, the restriction of the set-valued map F_{i+1} to this set is upper semicontinuous. Thus, by Lemma 19.3 it is measurable on X_ε^i. By the Luzin theorem, there exists such a closed set $X_\varepsilon^{i+1} \subset X_\varepsilon^i$ that $\mu(X_\varepsilon^i \setminus X_\varepsilon^{i+1}) \leq 2^{-(i+2)}\varepsilon$ and the restriction of F_{i+1} to the set X_ε^{i+1} is continuous. This completes the induction.

Set $X_\varepsilon = \bigcap_{i=1}^\infty X_\varepsilon^i$. The set X_ε is closed, $\mu(X \setminus X_\varepsilon) \leq \varepsilon$ and the restriction of each of the set-valued maps F_i to X_ε is continuous. Since $\varepsilon > 0$ is arbitrary, each of maps F_i is measurable on X. Set $\tilde{F}(x) = \bigcap_{i=1}^\infty F_i(x)$. For each $x \in X$, the set $\tilde{F}(x)$ is nonempty as the intersection of a decreasing sequence of compact sets. Thus, by Lemma 19.5 the set-valued map $\tilde{F}$ is measurable.

For all $x \in \tilde{X}$, where $\tilde{X} = \bigcup_{i=1}^\infty X_{1/i}$, the set $\tilde{F}(x)$ consists of a single point $\xi(x)$. This follows from the fact that if, for a given x, we have $\xi_1, \xi_2 \in \tilde{F}(x)$, then $\rho_Y(y_i, \xi_1) = \rho_Y(y_i, \xi_2)$ for all i, and hence $\xi_1 = \xi_2$, since, by construction, the set $\{y_1, y_2, \dots\}$ is everywhere dense in Y. It is obvious also that $\mu(X \setminus \tilde{X}) = 0$. So, the constructed map ξ is a measurable selection of a set-valued map F. □

Let us present a generalization of Theorem 19.6 which is frequently used in applications.

Theorem 19.7 (C. Castaing[5] [47]). *Let a set-valued map F be measurable. Then it admits a denumerable set of such measurable selections $\{\xi_1, \xi_2, \dots\}$, that for almost all $x \in X$ the set $\{\xi_1(x), \xi_2(x), \dots\}$ is everywhere dense in $F(x)$.*

Proof. Take the in Y everywhere dense subset $\{y_1, y_2, \dots\}$ and for all integers i, j set

$$X_{i,j} = \{x \in X \colon F(x) \bigcap B^Y(y_i, 1/j) \neq \emptyset\}.$$

By Theorem 19.2, each of the sets $X_{i,j}$ is measurable since $X_{i,j} = F^-(B^Y(y_i, 1/j))$. Define set-valued maps $F_{i,j} \colon X \to \mathcal{H}_c(Y)$ by the relation

$$F_{i,j}(x) = \begin{cases} F(x) \bigcap B^Y(y_i, 1/j), & x \in X_{i,j}, \\ F(x), & x \in X \setminus X_{i,j}. \end{cases}$$

By Theorem 19.2 each of these maps is measurable. This follows from the next evident formula which is true for an arbitrary closed set $A \subset Y$:

$$F^-(A) = \left(F^-(A) \bigcap (X \setminus X_{i,j})\right) \bigcup \left(F^-\left(B^Y(y_i, 1/j) \bigcap A\right) \bigcap X_{i,j}\right).$$

By Theorem 19.6 each of the set-valued maps $F_{i,j}$ admits a measurable selection $\xi_{i,j}$.

5 Charles Castaing (born 1932), a French mathematician.

Take arbitrary $x \in X$, $y \in F(x)$. For each integer j there exists an integer i such that $\rho_Y(y, y_i) \le 1/j$. Hence, $x \in X_{i,j}$ and therefore, $\xi_{i,j}(x) \in F_{i,j}(x) \subset B^Y(y_i, 1/j)$, yielding $\rho_Y(y, \xi_{i,j}(x)) \le 2/j$. This means that the denumerable set $\{ \xi_{i,j}(x) \}$ is everywhere dense in $F(x)$. ☐

Remark 19.8. In fact, the existence of such a denumerable set of measurable selections is equivalent to the measurability of the set-valued map F (see, e.g., [5], Ch. III). This set of selections is frequently called the Castaing representation of a set-valued map F. Theorems 19.6 and 19.7 are commonly referenced as the measurable choice theorems.

In conclusion of this section, consider the assertion which plays a fundamental role in the mathematical theory of control. In the literature it is often called *the Filippov*[6] *implicit function lemma* (see [48]).

Let Z be a complete metric space. Consider a measurable set-valued map $F\colon X \to \mathcal{H}_c(Y)$ and a map $G\colon X \times Y \to Z$ such that for each fixed $y \in Y$ the map $G(\cdot, y)$ is measurable and for every fixed $x \in X$ the map $G(x, \cdot)$ is continuous.

Theorem 19.9. *Let $g\colon X \to Z$ be a measurable map such that $g(x) \in G(x, F(x))$ for all $x \in X$. Then the set-valued map F admits such a measurable selection ξ that*

$$g(x) = G(x, \xi(x)) \ \forall x.$$

Proof. For $x \in X$, set

$$\tilde{F}(x) = \{ y \in F(x)\colon G(x, y) = g(x) \}.$$

It is evident that each of the sets $\tilde{F}(x)$ is compact. Let us show that the set-valued map $\tilde{F}$ is measurable. Indeed, take $\varepsilon > 0$. Applying to the map G the parametrized version of the Luzin theorem called *the Scorza Dragoni*[7] *theorem* (see, e.g., [21], Theorem 7.11) we can conclude that there exists such a closed subset $X_\varepsilon \subset X$, that $\mu(X \setminus X_\varepsilon) < \varepsilon$ and the restriction of G to $X_\varepsilon \times Y$ is continuous. By the Luzin theorem, we can assume, without loss of generality, that the restrictions of maps F and g to X_ε are also continuous.

Let us show that the restriction of the set-valued map $\tilde{F}$ to the set X_ε is upper semicontinuous. Indeed it is easy to verify (do it as an exercise!) that the restriction of $\tilde{F}$ to X_ε is closed. In addition, we have $\tilde{F} = \tilde{F} \bigcap F$, and moreover the restriction of F to X_ε is upper semicontinuous (it is even continuous). Applying Theorem 17.24 we see that the map $\tilde{F}$ on the set X_ε is upper semicontinuous and, hence, by Lemma 19.3 measurable. From the arbitrariness of $\varepsilon > 0$ we conclude that the set-valued map $\tilde{F}$ is measurable. By Theorem 19.7 it admits a measurable selection ξ which is obviously the desired one. ☐

---

6 Aleksei Fedorovich Filippov (1923–2006), a Russian mathematician.

7 Giuseppe Scorza Dragoni (1908–1996), an Italian mathematician.

20 The superposition set-valued operator

Let E, E_0 be separable Banach spaces; $I \subset \mathbb{R}$ a compact interval endowed with the Lebesgue measure μ.

Definition 20.1. A set-valued map $F: I \times E_0 \to \mathcal{H}_c(E)$ satisfies the upper Carathéodory conditions provided

(i) For every $x \in E_0$, the set-valued function $F(\cdot, x): I \to \mathcal{H}_c(E)$ is measurable;

(ii) For almost every $t \in I$, the set-valued map $F(t, \cdot): E_0 \to \mathcal{H}_c(E)$ is upper semicontinuous.

Definition 20.2. A set-valued map $F: I \times E_0 \to \mathcal{H}_c(E)$ satisfies the Carathéodory conditions if it satisfies condition (i) of the previous definition, but instead of (ii), the following condition holds true:

(ii′) For almost every $t \in I$, the set-valued map $F(t, \cdot): E_0 \to \mathcal{H}_c(E)$ is continuous.

It is clear that a set-valued map F satisfying the Carathéodory conditions can be considered a single-valued map into the metric space $\mathcal{H}_c(E)$ endowed with the Hausdorff metric h which obeys usual "single-valued" Carathéodory conditions. Therefore, we can conclude that the following Scorza Dragoni property holds true.

Theorem 20.3. *Let a set-valued map $F: I \times E_0 \to \mathcal{H}_c(E)$ satisfy the Carathéodory conditions. Then for every $\delta > 0$ there exists a closed subset $I_\delta \subset I$ such that $\mu(I \setminus I_\delta) \leq \delta$ and the restriction of F to $I_\delta \times E_0$ is continuous.*

Every set-valued map $F: I \times E_0 \to \mathcal{H}_c(E)$ generates the correspondence which assigns to each set-valued function Q from I to E_0 the set-valued function Φ from I to E defined by the formula

$$\Phi(t) = F(t, Q(t)).$$

Let us describe some properties of this correspondence.

Theorem 20.4. *If a set-valued map $F: I \times E_0 \to \mathcal{H}_c(E)$ satisfies the Carathéodory conditions then it is superpositionally measurable in the sense that for every measurable multifunction $Q: I \to \mathcal{H}_c(E_0)$ the set-valued function Φ is measurable.*

Proof. Without loss of generality, we can assume that the set-valued map F is continuous in the second argument for all $t \in I$. Then, by Proposition 17.30, the set-valued function Φ has compact values.

Fix an arbitrary $\delta > 0$ and choose a closed set $I_\delta \subset I$ such that $\mu(I \setminus I_\delta) \leq \delta$ and the restrictions of Q to I_δ and F to $I_\delta \times E_0$ are continuous (the Luzin theorem and Theorem 20.3). Then it is easy to see (verify this as an exercise!) that the restriction of the multifunction Φ to I_δ is also continuous. We conclude the proof by the application of the Luzin theorem. □

DOI 10.1515/9783110460308-021

Following the same line of reasoning, we can justify the next conditions for superpositional measurability.

Theorem 20.5. *If a set-valued map $F\colon I \times E_0 \to \mathcal{H}_c(E)$ is upper or lower semicontinuous, then it is superpositionally measurable.*

Notice that the property of superpositional measurability disappears if we weaken the Carathéodory conditions to the upper ones. Consider the following example (see [17], Example 1.5.20).

Example 20.6. Let $D \subset [0,1]$ be an arbitrary nonmeasurable set. Consider the function $\varphi\colon [0,1] \to \mathbb{R}$,

$$\varphi(t) = \begin{cases} t, & t \in [0,1] \setminus D, \\ t+1, & t \in D. \end{cases}$$

Define the set-valued map $F\colon [0,1] \times \mathbb{R} \to \mathcal{H}_c(\mathbb{R})$,

$$F(t,x) = \begin{cases} \{1\}, & (t,x) \in ([0,1] \times \mathbb{R}) \setminus G_\varphi, \\ [0,1], & (t,x) \in G_\varphi, \end{cases}$$

where G_φ is the graph of φ. It is easy to see that the set-valued map F satisfies the upper Carathéodory conditions (but not the Carathéodory conditions).

If now $Q = i\colon [0,1] \to \mathbb{R}$ is the inclusion map then the set-valued map $\Phi\colon [0,1] \to \mathcal{H}_c(\mathbb{R})$,

$$\Phi(t) = F(t,t) = \begin{cases} [0,1], & t \in [0,1] \setminus D, \\ \{1\}, & t \in D \end{cases}$$

is not measurable.

This example shows also that the Scorza Dragoni property (see Theorem 20.3) does not hold for upper Carathéodory set-valued maps. Indeed, it is not difficult to verify (do it as an exercise!) that every set-valued map satisfying the Scorza Dragoni property is necessarily superpositionally measurable.

Nevertheless, if F is the upper Carathéodory set-valued map then the resulting set-valued function Φ possesses some "good" properties. In particular, the following is true.

Theorem 20.7. *If a set-valued map $F\colon I \times E_0 \to \mathcal{H}_c(E)$ satisfies the upper Carathéodory conditions then it is superpositionally selectionable in the sense that for each measurable set-valued function $Q\colon I \to \mathcal{H}_c(E_0)$ there exists a measurable selection $\varphi\colon I \to E$ of the set-valued function $\Phi\colon I \to \mathcal{H}_c(E)$,*

$$\Phi(t) = F(t, Q(t)).$$

Proof. Taking an arbitrary measurable selection $q\colon I \to E_0$ of the set-valued function Q we can assume, without loss of generality, that the set-valued function Φ has the form $\Phi(t) = F(t, q(t))$. Let $\{q_n\}_{n=1}^{\infty}$, $q_n : I \to E_0$ be a sequence of step functions converging almost everywhere on I to q. Using condition (i) of Definition 20.1 we can form a sequence of measurable selections $\{\varphi_n\}_{n=1}^{\infty}$, $\varphi_n : I \to E$,

$$\varphi_n(t) \in F(t, q_n(t)) \quad \text{a.e. } t \in I.$$

Now for a.e. $t \in I$ and $m \geq 1$ define

$$\Phi_m(t) = \overline{\bigcup_{k=m}^{\infty} \varphi_k(t)}.$$

From condition (ii) of Definition 20.1 and Proposition 17.30 it follows that the sets $\Phi_m(t)$ are compact and moreover (see Remark 19.8) $\{\Phi_m\}_{m=1}^{\infty}$, $\Phi_m : I \to \mathcal{H}_c(E)$ is the sequence of set-valued functions.

For a.e. $t \in I$ the sets $\Phi_m(t)$, $m \geq 1$ form the decreasing sequence of compact sets and taking into account Lemma 19.5 we can conclude that the set-valued function $\tilde{\Phi} : I \to \mathcal{H}_c(E)$,

$$\tilde{\Phi}(t) = \bigcap_{m=1}^{\infty} \Phi_m(t) \quad \text{for a.e. } t \in I$$

is well defined and measurable. From condition (ii) it follows that

$$\tilde{\Phi}(t) \subseteq \Phi(t) = F(t, q(t))$$

for a.e. $t \in I$ and, applying Theorem 19.6, we get a measurable selection φ of $\tilde{\Phi}$ which is clearly the desired one. $\qquad\square$

Now introduce the following notion.

Definition 20.8. We will say that an upper Carathéodory set-valued map $F : I \times E_0 \to \mathcal{H}_c(E)$ satisfies the L^1-upper Carathéodory conditions if it obeys the following condition of local integral boundedness:

(iii) for each $r > 0$ there exists such a function $v_r \in L^1_+(I)$ that

$$\|F(t, x)\| : = \max \{ \|y\| : y \in F(t, x) \} \leq v_r(t) \quad \text{for a.e. } t \in I$$

for all x, $\|x\| \leq r$.

From Theorem 20.7 and the above definition it easily follows that each L^1-upper Carathéodory set-valued map $F : I \times E_0 \to \mathcal{H}_c(E)$ generates the set-valued map $\mathcal{P}_F$ from the space of continuous functions $C(I; E_0)$ into the space of Bochner[1] integrable functions $L^1(I; E)$ assigning to each continuous function $q \in C(I; E_0)$ the set of all integrable selections of the multifunction Φ, $\Phi(t) = F(t, q(t))$. We will call $\mathcal{P}_F$ the superposition set-valued operator.

---

1 Salomon Bochner (1899–1982), an American mathematician.

We will study the continuity properties of this operator. To this end, let us present a few auxiliary lemmas. The next assertion follows from the criterion of the weak compactness in the space of integrable functions (see [49]).

Lemma 20.9. *Suppose that a sequence of functions $\{f_n\}_{n=1}^{\infty} \subset L^1(I; E)$ is such that:*
(1) *there exists an integrable function $v \in L_+^1(I)$ such that*

$$\|f_n(t)\| \le v(t) \quad \text{for a.e. } t \in I$$

for all $n \ge 1$;
(2) *the set $\{f_n(t)\}_{n=1}^{\infty}$ is relatively compact in E for a.e. $t \in I$.*
Then the sequence of functions $\{f_n\}_{n=1}^{\infty}$ is weakly compact, i.e., a weakly convergent subsequence can be extracted from it.

Lemma 20.10 (of Mazur). *Let $\{u_n\}_{n=1}^{\infty}$ be a sequence of elements of a normed space weakly convergent to u. Then there exists a double sequence of nonnegative numbers $\{\lambda_{ik}\}_{i=1\,k=1}^{\infty\ \ \infty}$ such that:*
(a) *$\sum_{k=i}^{\infty} \lambda_{ik} = 1$ for all $i \ge 1$;*
(b) *for each $i \ge 1$ there exists a number $k_0(i)$ such that $\lambda_{ik} = 0$ for all $k \ge k_0(i)$;*
(c) *the sequence of convex combinations $\{\tilde{u}_i\}_{i=1}^{\infty}$,*

$$\tilde{u}_i = \sum_{k=i}^{\infty} \lambda_{ik} u_k$$

converges to u with respect to the norm.

The proof can be found, for example, in [6].

Lemma 20.11 (see [31], Ch. IV, Theorem 38). *If a sequence of functions $\{f_n\}_{n=1}^{\infty} \subset L^1(I; E)$ converges to a function f with respect to the norm of the space $L^1(I; E)$ then there exists a subsequence $\{f_{n_i}\}$ which converges to f almost everywhere on I.*

Under the additional assumption that the generating set-valued map F has convex values, the following assertion which is due to A. Lasota[2] and Z. Opial[3] about the closedness of the superposition set-valued operator holds true.

Denote by $\mathcal{H}_{cv}$ and $\mathcal{H}_{clv}$ collections consisting of nonempty compact convex and, respectively, closed convex subsets of a corresponding space.

Theorem 20.12. *Suppose that a set-valued map $F: I \times E_0 \to \mathcal{H}_{cv}(E)$ satisfies the L^1-upper Carathéodory conditions and $A: L^1(I; E) \to E_1$, where E_1 is a normed space, is a bounded linear operator. Then the composition*

$$A \circ \mathcal{P}_F: C(I; E_0) \to \mathcal{H}_{clv}(E_1)$$

is a closed set-valued map.

---

2 Andrzej Lasota (1932–2006), a Polish mathematician.
3 Zdzislaw Opial (1930–1974), a Polish mathematician.

Proof. First of all, notice that the convexity of values of the set-valued map $A \circ \mathcal{P}_F$ follows from the convexity of values of F and the linearity of the operator A.

Consider sequences $\{q_n\}_{n=1}^{\infty}$, $q_n \in C(I; E_0$, $\{z_n\}_{n=1}^{\infty}$, $z_n \in E_1$ such that

$$\lim_{n\to\infty} \|q_n - q\|_C = 0, \quad z_n \in A \circ \mathcal{P}_F(q_n), \quad \lim_{n\to\infty} \|z_n - z\|_{E_1} = 0.$$

Choose a sequence $\{f_n\}_{n=1}^{\infty} \subset L^1(I; E)$, $f_n \in \mathcal{P}_F(q_n)$, $z_n = A(f_n)$, $n \geq 1$. Taking into account the convergence of the sequence $\{q_n\}_{n=1}^{\infty}$ and applying Proposition 17.30 it is easy to verify that the sequence $\{f_n\}_{n=1}^{\infty}$ satisfies conditions (1) and (2) of Lemma 20.9 and hence it is weakly compact. Passing to a subsequence, if necessary, we will assume, without loss of generality, that it weakly converges to a function $f \in L^1(I; E)$.

By Lemma 20.10 there exists a sequence $\{\tilde{f}_i\}_{i=1}^{\infty} \subset L^1(I; E)$, $\tilde{f}_i = \sum_{k=i}^{\infty} \lambda_{ik} f_k$ which converges to f with respect to the norm of the space $L^1(I; E)$. Applying Lemma 20.11 we again can assume without loss of generality that the sequence $\{\tilde{f}_i\}_{i=1}^{\infty}$ converges to f almost everywhere on I.

From condition (ii) of Definition 20.1 it follows that, given a positive $\varepsilon > 0$, for almost every $t \in I$ there exists a number $i_0 = i_0(\varepsilon, t)$ such that

$$F(t, q_i(t)) \subset O(F(t, q(t)), \varepsilon) \quad \text{for } i \geq i_0,$$

where $O(F(t, q(t)), \varepsilon)$ denotes the ε-neighborhood of the set $F(t, q(t))$. But then also

$$f_i(t) \in O(F(t, q(t)), \varepsilon) \quad \text{for } i \geq i_0,$$

and hence, by convexity of $O(F(t, q(t)), \varepsilon)$, we get

$$\tilde{f}_i(t) \in O(F(t, q(t)), \varepsilon) \quad \text{for } i \geq i_0,$$

implying, by the arbitrariness of ε,

$$f(t) \in F(t, q(t)) \quad \text{for a.e. } t \in I,$$

i.e., $f \in \mathcal{P}_F(q)$.

On the other side,

$$A(\tilde{f}_i) = \sum_{k=i}^{\infty} \lambda_{ik} A(f_k) = \sum_{k=i}^{\infty} \lambda_{ik} z_k,$$

implying $\lim_{i\to\infty} \|A(\tilde{f}_i) - z\|_{E_1} = 0$.

From the continuity of the operator A it follows that $z = A(f)$, yielding $z \in A \circ \mathcal{P}_F(q)$ that concludes the proof. □

Consider the following important particular case. Let $I = [t_0, T]$ and $j\colon L^1(I; E) \to C(I; E)$ be the integral operator

$$j(f)(t) = \int_{t_0}^{t} f(s)\, ds.$$

Definition 20.13. The composition

$$j \circ \mathcal{P}_F \colon C(I; E_0) \to \mathcal{H}_{clv}(C(I; E))$$

is called *the integral set-valued operator generated by the set-valued map F.*

Corollary 20.14. *Suppose that a set-valued map $F \colon I \times E_0 \to \mathcal{H}_{cv}(E)$ satisfies the L^1-upper Carathéodory conditions. Then the integral set-valued operator $j \circ \mathcal{P}_F$ is a closed set-valued map.*

21 The Michael theorem and continuous selections. Lipschitz selections. Single-valued approximations

In this section we consider various conditions providing that a given set-valued map admits a continuous selection and moreover, a Lipschitz continuous selection. We will start with the classical continuous selection result which is due to E. Michael[1] (see [50]).

Everywhere in this section (with the exception of the last theorem) we will assume that (X, ρ) is a metric space and Y is a Banach space. The lower semicontinuity of a set-valued map will be understood in the sense of Definition 17.28. Recall that it is equivalent to sequential lower semicontinuity (see Chapter 17).

Theorem 21.1. *Each lower semicontinuous set-valued map $F: X \to \mathcal{H}_{\text{clv}}(Y)$ admits a continuous selection.*

Proof. In the general case, the proof of the Michael theorem is fairly uneasy and needs some additional techniques (for example, the lemma on the partition of unity is necessary). The complete proof can be found, for example, in [13], § 2.9 or in [17], § 1.4. We will present here the proof under additional assumptions that the set-valued map F is sequentially continuous (i.e., it is both sequentially upper semicontinuous and lower semicontinuous) and the space Y is Hilbert. (These assumptions essentially simplify the reasoning).

Thus, let the set-valued map F be continuous and Y a Hilbert space. For each fixed $x \in X$ consider the minimization problem

$$|y|^2 \to \min, \quad y \in F(x). \tag{21.1}$$

The minimum in this problem is achieved since $F(x)$ is a closed convex subset in a Hilbert space (see, e.g., [39], Vol. 2, Ch. 8, § 2, Theorem 13). Moreover, this minimum is achieved in a unique point. This follows from the fact that if $y_1, y_2 \in F(x)$, $y_1 \neq y_2$ and $|y_1| = |y_2| = r$, then $(y_1 + y_2)/2 \in F(x)$, since the set $F(x)$ is convex and $|(y_1 + y_2)/2| < r$.

Denote by $f(x) \in F(x)$ at which the minimum in Theorem (21.1) is achieved. We need to show that the selection f is continuous.

Take an arbitrary point $x_0 \in X$ and set $y_0 = f(x_0)$, $D = F(x_0)$. Let us prove that

$$|y|^2 \geq |y_0|^2 + |y - y_0|^2 \quad \forall y \in D. \tag{21.2}$$

To this end, first show that $\langle y_0, y - y_0\rangle \geq 0$ $\forall y \in D$. Indeed, for $\theta \in [0, 1]$ set $y(\theta) = y_0 + \theta(y - y_0)$. Then $y(\theta) \in D$ $\forall \theta \in [0, 1]$, since the set D is convex. So, $|y(\theta)|^2 \geq |y_0|^2$ $\forall \theta \in$

1 Ernest Michael (1925–2013), an American mathematician.

DOI 10.1515/9783110460308-022

$[0, 1]$, since y_0 is the solution of problem (21.1) for $x = x_0$, that yields $2\theta\langle y_0, y - y_0\rangle + \theta^2|y - y_0|^2 \geq 0$. Dividing this inequality over $\theta > 0$, and tending $\theta \to 0+$ we get the desired inequality. By virtue of the obtained inequality we have for $y \in D$

$$|y|^2 = |y_0 + (y - y_0)|^2 = |y_0|^2 + 2\langle y_0, y - y_0\rangle + |y - y_0|^2 \geq |y_0|^2 + |y - y_0|^2,$$

proving (21.2).

Suppose now that the map f is not continuous at the point x_0. Then there exist such an $\varepsilon > 0$ and a sequence $\{x_n\}$ convergent to x_0, that $|f(x_n) - y_0|^2 \geq 3\varepsilon\ \forall n$. But $f(x_n) \in F(x_n)\ \forall n$. From the continuity of the set-valued map F it follows that it is sequentially upper semicontinuous. Hence, $\operatorname{dist}(f(x_n), F(x_0)) \to 0$ and thus there exist such $y_n \in F(x_0)$, that $|y_n - f(x_n)| \to 0$, $n \to \infty$. Therefore $|y_n - y_0|^2 \geq 2\varepsilon$ for all sufficiently large n, yielding, by (21.2) $|y_n|^2 \geq |y_0|^2 + 2\varepsilon$. From here it follows that $|f(x_n)|^2 \geq |y_0|^2 + \varepsilon$ for all sufficiently large n.

On the other hand, by the sequential lower semicontinuity of the set-valued map F there exists such a sequence $\{\tilde{y}_n\}$, that $\tilde{y}_n \to y_0$ and $\tilde{y}_n \in F(x_n)\ \forall n$. Hence $|f(x_n)|^2 \leq |\tilde{y}_n|^2\ \forall n$ and therefore $|f(x_n)|^2 \leq |y_0|^2 + \varepsilon/2$ for all sufficiently large n. The obtained contradiction with the above inequality demonstrates the continuity of f at the point x_0 and concludes the proof. □

The Michael theorem can be generalized to the following result on the extension of a continuous selection.

Theorem 21.2. *Let $F\colon X \to \mathcal{H}_{clv}(Y)$ be a lower semicontinuous set-valued map, $C \subset X$ a closed subset and $g\colon C \to Y$ a continuous map such that $g(x) \in F(x)\ \forall x \in C$. Then F admits a continuous selection $f\colon X \to Y$ such that $f(x) = g(x)\ \forall x \in C$.*

Proof. Define the set-valued map $\tilde{F}\colon X \to \mathcal{H}_{clv}(Y)$ by the relation

$$\tilde{F}(x) = \begin{cases} F(x), & x \notin C \\ \{g(x)\}, & x \in C. \end{cases}$$

Let us show that the set-valued map $\tilde{F}$ is lower semicontinuous. In fact, take an arbitrary $x_0 \in X$. If $x_0 \notin C$ then, since the set C is closed, there exists $\varepsilon > 0$ such that $O^X(x_0, \varepsilon) \bigcap C = \emptyset$ and hence $\tilde{F}(x) = F(x)$ for all $x \in O^X(x_0, \varepsilon)$, that immediately implies that $\tilde{F}$ is lower semicontinuous at the point x_0.

Now, let $x_0 \in C$. Set $y_0 = g(x_0) = \tilde{F}(x_0)$. Suppose that $\tilde{F}$ is not sequentially lower semicontinuous at the point x_0. Then there exist $\varepsilon > 0$ and a sequence $\{x_n\}$ convergent to x_0 such that $\operatorname{dist}(y_0, \tilde{F}(x_n)) \geq \varepsilon$ for all n. Consider two cases. Let, first, $x_n \in C$ for infinitely many numbers n. Then, by definition, $\tilde{F}(x_n) = \{g(x_n)\}$ for these numbers n, that leads to the contradiction since by the continuity of g we have $g(x_n) \to g(x_0) = y_0$. In the second case, i.e., when $x_n \notin C$ for all sufficiently large numbers n, the desired contradiction follows immediately from the sequential lower semicontinuity of the initial map F.

So, it is proved that the constructed set-valued map $\tilde{F}$ is lower semicontinuous. Applying to it the Michael theorem, we obtain the desired result. □

Corollary 21.3. *Let $F\colon X \to \mathcal{H}_{clv}(Y)$ be a lower semicontinuous set-valued map, $x_0 \in X$ and $y_0 \in F(x_0)$ arbitrary points. Then F admits a continuous selection $f\colon X \to Y$ such that $f(x_0) = y_0$.*

Notice that similarly to measurable set-valued maps, there exists a dense "tube" of continuous selections for a lower semicontinuous set-valued map. Namely, the following assertion holds true.

Theorem 21.4. *Let $F\colon X \to \mathcal{H}_{clv}(Y)$ be a lower semicontinuous set-valued map, where Y is a separable Banach space. Then F admits a countable family of continuous selections $\{f_n\}$, such that for each $x \in X$ the sequence $\{f_n(x)\}$ is everywhere dense in the set $F(x)$.*

Proof. The proof of this theorem in general case may be found in [51]. We will give here the proof (suggested by B. D. Gel'man) under additional assumptions that the set-valued map F is continuous and Y is a Hilbert space.

Choose in Y a denumerable everywhere dense subset $\{y_1, y_2, \dots\}$. For each fixed number n and $x \in X$ consider the minimization problem

$$|y - y_n|^2 \to \min, \quad y \in F(x).$$

From the proof of the Michael theorem we remember that the solution of this problem exists and is unique and, moreover, if we denote this solution by $f_n(x)$ then the map f_n is a continuous selection of the map F.

Fix arbitrary $x \in X$. We show that the set $\{f_1(x), f_2(x), \dots\}$ is everywhere dense in the set $F(x)$. Indeed, take arbitrary $y \in F(x)$ and $\varepsilon > 0$. Choose such a number m, that $|y_m - y| \le \varepsilon/2$. Then $\operatorname{dist}(y_m, F(x)) \le \varepsilon/2$ and, hence by construction $|f_m(x) - y_m| \le \varepsilon/2$ and the triangle inequality yields $|f_m(x) - y| \le \varepsilon$ proving the desired. □

The assumption about the convexity of all values $F(x)$, $x \in X$ in the Michael theorem can be slightly weakened if we suppose that convexity is violated only on "a very small" (in a topological sense) set $Z \subset X$ (details can be found in [52] and in the next section). However, the convexity assumption cannot be completely omitted even if the map F is supposed to be continuous. This is demonstrated by the following example (see [17], Example 1.4.6).

Example 21.5. Let $D \subset \mathbb{R}^n$ be a closed unit disc centered at the origin, S its boundary and F a set-valued map defined on D in the following way:

$$F(x) = \begin{cases} S \setminus D(x), & x \neq 0; \\ S, & x = 0; \end{cases}$$

where $D(x)$ denotes the disc of radius $\|x\|$ centered at $\frac{x}{\|x\|}$. Suppose that F admits a continuous selection $f\colon D \to D$. Then, by the Brouwer fixed point theorem (see Theorem 26.5 below), there exists a point $x_0 \in D$ such that $x_0 = f(x_0)$ that yields $x_0 \in F(x_0)$, that is impossible by the definition of F.

Now consider the problem of existence of Lipschitz selections.

Theorem 21.6. *Suppose that $Y = \mathbb{R}^n$; a set-valued map F has convex compact values and satisfies the Lipschitz condition with a constant L, i.e.,*

$$h(F(x_1), F(x_2)) \leq L\rho_X(x_1, x_2) \quad \forall x_1, x_2 \in X.$$

Then F admits a selection f satisfying the Lipschitz condition with the constant $\tilde{L} = LL_n$. Here

$$L_n = \frac{2}{\sqrt{\pi}} \frac{\Gamma(\frac{n}{2}+1)}{\Gamma\left(\frac{n+1}{2}\right)},$$

where $\Gamma(\alpha) = \int_0^\infty t^{\alpha-1} \exp(-t)\, dt$ is the Euler[2] gamma function.

Proof. For an arbitrary convex compact $K \subset \mathbb{R}^n$, its Steiner[3] center $s(K)$ is defined by the formula

$$s(K) = \frac{1}{v_n} \int_{S_n} c(p, K) p\, dp.$$

Here S_n is the unit sphere in $\mathbb{R}^n$, v_n is the value of the unit ball in $\mathbb{R}^n$, and $c(\cdot, K)$ is the support function of the set K. It is known (see [13], Ch. 2, § 2.1, lemmas 2.1.3 and 2.1.4), that $s(K) \in K$ and $|s(K_1) - s(K_2)| \leq L_n h(K_1, K_2)$ for arbitrary convex compact sets K, K_1, K_2 from $\mathbb{R}^n$. For $x \in X$, set $f(x) = s(F(x))$. Then, obviously, $f(x) \in F(x)$ for all $x \in X$ and $|f(x_1) - f(x_2)| = |s(F(x_1)) - s(F(x_2))| \leq L_n h(F(x_1), F(x_2)) \leq L_n L \rho_X(x_1, x_2)$ for arbitrary $x_1, x_2 \in X$. Thus, the constructed selection f is desired. □

Notice that "the universal constant" L_n depends only on the dimension n, satisfies the estimate $L_n \leq \sqrt{n}$ and behaves under the growth of n approximately as $\sqrt{n}$.

Remark 21.7. The assumption in the above theorem that the space Y is finite-dimensional is essential. There exists an example (see [53]) demonstrating the absence of a Lipschitz selection for a Lipschitz set-valued map in cases when Y is infinite-dimensional.

It is worth noting that, as the following example demonstrates, there exist Lipschitz set-valued maps which do not posses even continuous selections (for details see [54]).

Example 21.8. Consider the map $f\colon \mathbb{R}^2 \to \mathbb{R}^2$ defined by the formula

$$f(x) = \frac{1}{|x|}(x_1^2 - x_2^2, 2x_1x_2) \quad \forall x \neq 0,\ f(0) = 0,$$

2 Leonhard Euler (1707–1783), a Swiss, German and Russian mathematician and mechanician.
3 Jacob Steiner (1796–1863), a Swiss mathematician.

where $x = (x_1, x_2)$, $|x| = \sqrt{x_1^2 + x_2^2}$. Define by F the set-valued map inverse to f, i.e.,

$$F(y) = \{x \in \mathbb{R}^2 : y = f(x)\}.$$

From the continuity of f it follows that F has closed values. Let us show that the set $F(y)$ is nonempty for each $y \in \mathbb{R}^2$ and the set-valued map F is 1-Lipschitz.

To do so, it is sufficient to notice that for the map $\varphi : \mathbb{C} \to \mathbb{C}$, defined as

$$\varphi(z) = \begin{cases} z^2/|z|, & \text{if } z \neq 0, \\ 0, & \text{if } z = 0, \end{cases}$$

the following relation is true:

$$\forall z_0, w \in \mathbb{C}\ \exists z \in \mathbb{C} : w = \varphi(z) \text{ and } |z - z_0| \leq |w - \varphi(z_0)|, \tag{21.3}$$

where $\mathbb{C}$ is the field of complex numbers.

In the case when $z_0 \neq 0$, $w = 0$, the point $z = 0$ obviously satisfies (21.3). If $z_0 = 0$, then each solution z of the equation $\varphi(z) = w$ satisfies (21.3).

So, F has a nonempty closed values and is 1-Lipschitz. But each even map in $\mathbb{R}^n$ has no continuous right inverse (see Lemma 1 in [54]), so F does not admit a continuous selection.

In the Michael theorem the assumption on the lower semicontinuity of a set-valued map can neither be omitted nor replaced with the upper semicontinuity supposition, that demonstrates the map F_2 from Exercise 17.4. It is evident that each single-valued selection of this set-valued map should have the discontinuity at the point $x = 1$. In case we wish to study upper semicontinuous set-valued maps by means of single-valued maps, we can use single-valued approximations.

To this end, we will need the following auxiliary assertion which is called *the Lebesgue covering lemma* (see, e.g., [55]).

Lemma 21.9. *Let $\mathcal{U}$ be an open covering of a compact subset $\mathcal{K}$ of a metric space. Then there exists a number $\delta > 0$ (called the Lebesgue number of $\mathcal{U}$) such that an open ball of radius δ with the center at an arbitrary point of $\mathcal{K}$ is contained in a certain element of the covering $\mathcal{U}$.*

The following theorem on the existence of *a single-valued ε-approximation* holds true.[4]

Theorem 21.10. *Let Y be a normed space. Then for each upper semicontinuous set-valued map $F : X \to \mathcal{H}_{\mathrm{clv}}(Y)$ and every $\varepsilon > 0$ there exists a continuous map $f_\varepsilon > 0$ such that:*

(i) *for each $x \in X$ there exists $x' \in X$ such that $\rho_X(x, x') < \varepsilon$ and*

$$f_\varepsilon(x) \cup F(x) \subset O(F(x'), \varepsilon);$$

(ii) $f_\varepsilon(X) \subset \operatorname{conv} F(X)$.

4 Various versions of this assertion can be found in the works [56, 57, 58, 59].

Proof. The complete proof of this result also needs some topological techniques, including the partition of unity (see, e.g., [17], Theorem 1.4.11). To simplify the proof, we will assume that X may be presented as a finite polyhedron K, in the sense that it has the form of a union of a finite number of simplexes such that the intersection of each different two of them is either empty or is their joint face.[5]

Since the map F is upper semicontinuous, for each point $x \in X$ the set

$$U_x = \{\tilde{x} \in X : F(\tilde{x}) \subset O(F(x), \varepsilon)\}$$

is open. Let $\delta > 0$ be the Lebesgue number of the covering $\{U_x\}_{x \in X}$ of the space X. Suppose that the polyhedron K is triangulated so finely that $\operatorname{diam} \sigma < \delta$ for all simplexes σ from K.

For each vertex x_i of every simplex σ take an arbitrary $y_i \in F(x_i)$. Now, let a point $x \in X$ belong to a simplex $\sigma \in K$ with the vertices $x_0, \dots, x_n$. Then if $(\alpha_0, \dots, \alpha_n)$ are the barycentric coordinates of x in σ, define

$$f_\varepsilon(x) = \alpha_0 y_0 + \cdots + \alpha_n y_n.$$

It is easy to see that defining f_ε in this way on each simplex $\sigma \in K$ we get the desired ε-approximation of F. □

If we will introduce the metric ρ in the Cartesian product $X \times Y$ through the metrics ρ_X of the space X and norm of the space Y by the formula

$$\rho((x, y), (x', y')) = \max\{\rho_X(x, x'), \|y - y'\|\}$$

then we obtain the geometrically clear interpretation of a single-valued ε-approximation: the graph $\operatorname{gph}(f_\varepsilon)$ is contained in the ε-neighborhood of the graph $\operatorname{gph}(F)$, in other words, the deviation of $\operatorname{gph}(f_\varepsilon)$ from $\operatorname{gph}(F)$ is less than ε.

The proved approximation theorem is important not only by itself, but it can also be used as a convenient tool for various constructions, e.g., below we will apply it to build the topological degree for set-valued maps.

Remark 21.11. A single-valued ε-approximation satisfying condition (ii) of Theorem 21.10 will be called regular.

5 Surely, a polyhedron K may be considered a "curvilinear" one, i.e., each simplex $\sigma \in K$ may be treated as a homeomorphic image of a usual, "rectilinear" simplex. Under this approach, for example, a sphere in $\mathbb{R}^3$ may be triangulated into a polyhedron with the help of the equator and a few meridians.

22 Special selections of set-valued maps

Above it was shown that each measurable set-valued map defined on a metric space X with measure admits a measurable selection, then it admits a measurable selection (Theorem 19.6) and if it is lower semicontinuous and has closed convex values then it has a continuous selection (Theorem 21.1).

Comparing these two results, we see that the continuous selection theorem demands, generally speaking, more assumptions but at the same time it yields more. In this connection, the natural question arises: if we suppose that a measurable set-valued map also satisfies on a certain subset $\tilde{X} \subseteq X$ the conditions of the Michael theorem, can we guarantee the existence of a measurable selection which should be continuous in each point $x \in \tilde{X}$? Following the paper [60], we will consider this question in the present section.[1]

We will assume that on a metric space (X, ρ) a denumerably additive regular and nonnegative measure μ with $\mu(X)$ finite is defined. Let Y be a separable Banach space[2] and $F\colon X \to \mathcal{H}_c(Y)$ a set-valued map.

First let us present a generalization of the Michael theorem mentioned in the previous section. To this end, introduce the notion of a set whose topological dimension (introduced by coverings) equals zero.

Let B be a nonempty subspace of a metric space X. We will say that the set B is *zero-dimensional* (it will be written as $\dim B = 0$), if an arbitrary finite open covering $O_1, \dots, O_m$ of B admits a finite open covering $V_1, \dots, V_l$ of B which is refined in it and such that $V_i \bigcap V_j = \emptyset \ \forall i \neq j$.[3]

Theorem 22.1. *Let B be a zero-dimensional subset of the metric space X. Suppose that the set-valued map F is lower semicontinuous and for each $x \in X \setminus B$ the set $F(x)$ is convex. Then F admits a continuous selection.*

This assertion is proved in [52].

Now, let us consider the main results of this section.

Theorem 22.2. *Let a set-valued map F be measurable and the following sets are given: a closed set $B_0 \subset X$ which is either empty or zero-dimensional and a set $B_1 \subseteq X$ such that the sets $F(x)$ are convex for each $x \in B_1$. Assume that F is lower semicontinuous at*

1 During an initial reading, this section may be omitted.

2 In this section, in notation for the metric ρ in the space X, for convenience, we will omit the lower index $_X$ since in the Banach space Y we will use the norm instead of the metric.

3 Recall that the covering $V_1, \dots, V_l$ is said to be refined into the covering $O_1, \dots, O_m$, if for each $i \leq l$ there exists such a $j \leq m$, that $V_i \subset O_j$. More details about the dimension of a subset of a metric space can be found in [61], Ch. 2.

DOI 10.1515/9783110460308-023

each point $x \in B_0 \bigcup \mathrm{cl}\, B_1$. *Then* F *admits a measurable selection* f, *which is continuous at each point* $x \in B = B_0 \bigcup B_1$.

Before proving this assertion, let us pay attention to the following. Let a map $f: X \to Y$ and a set $C \subset X$ be given and $f|_C$ be the restriction of this map to C. Then the continuity of this restriction is not the same as the continuity of the initial map in all points of the set C. Indeed, the continuity of the restriction $f|_C$ means that if $x_0 \in C$ and $\{x_n\} \subset C$, $x_n \to x_0$, then $f(x_n) \to f(x_0)$. At the same time, the continuity of the map f in all points of the set C means that if $x_0 \in C$ and $x_n \to x_0$, then $f(x_n) \to f(x_0)$. Thus, the latter implies the continuity of the restriction $f|_C$, but, naturally, not conversely.

The proof of Theorem 22.2 is based on the following assertion.

Lemma 22.3. *Let* $C \subseteq X$ *be a closed set. Suppose that the set-valued map* F *is lower semicontinuous at each point* $x \in C$ *and its restriction* $F|_C$ *admits a continuous selection* $\tilde{f}$ *on* C. *Then* F *has a continuous selection* f *such that* $\tilde{f} = f|_C$ *and* f *is continuous at each point* $x \in C$.

Proof. Denote by

$$\Gamma = \{(x, \tilde{f}(x)) : x \in C\} \subset X \times Y$$

the graph of the map $\tilde{f}$. For an arbitrary $x \in X$ set

$$d(x) = \{\inf \|y - \tilde{f}(\xi)\| + \rho(x, \xi) \mid y \in F(x),\ \xi \in C\},$$

i.e., $d(x)$ is the distance between the graphs Γ and gph F. Obviously, $d(x) = 0 \Longleftrightarrow x \in C$. For $x \notin C$ set

$$\hat{F}(x) = \{y \in F(x) : \exists \xi \in C,\ \|y - \tilde{f}(\xi)\| + \rho(x, \xi) < 2d(x)\}.$$

Let us show that so defined on $X \setminus C$ the set-valued map $\hat{F}$ admits a measurable selection $\hat{f}: (X \setminus C) \to Y$. To this end, first prove that the function d is measurable. In fact, by the measurable choice Theorem 19.7 the set-valued map F admits such a denumerable set of measurable selections $\{f_n\}$, that for almost every $x \in X$ the set $\{f_1(x), f_2(x), \dots\}$ is everywhere dense in $F(x)$. Then, obviously, the following relation holds true:

$$d(x) = \inf_i (\inf\{\|f_i(x) - \tilde{f}(\xi)\| + \rho(x, \xi),\ \xi \in C\}). \tag{22.1}$$

Fix an arbitrary number i and prove that the function

$$y_i(x) = \inf\{\|f_i(x) - \tilde{f}(\xi)\| + \rho(x, \xi),\ \xi \in C\}$$

is measurable. Indeed, by the Luzin theorem there exists such a sequence of closed subsets $\{X_n\}$ of X, that $\mu(X \setminus X_n) \le n^{-1}$ and the restriction of the map f_i on each set X_n is continuous. It is easy to see that the restriction of the function y_i on each set X_n is also continuous implying, by the Luzin theorem, the measurability of the function y_i itself.

It is well known (see, e.g., [27]) that a pointwise limit of a sequence of measurable functions is measurable and the minimum of a finite number of measurable functions is measurable and hence the greatest lower bound of a sequence of measurable functions is measurable. From here, by the above proved measurability of functions y_i, from representation (22.1) it follows that the function d is measurable.

Apply again the Luzin theorem. It yields that there exists such a sequence of closed sets $\{X_n\}$, that $\mu(X \setminus X_n) \le n^{-1}$ and for each n the restrictions of the function d and each map f_i on the set X_n are continuous. Set $\hat{X}_n = X_n \setminus C$. Then, by construction $d(x) > 0$ for all $x \in \hat{X}_n$ and all n.

Let us turn to the construction of the map $\hat{f}$. First let us define it on the set $\hat{X}_1$. To do so, for each positive integer i set

$$Z_i = \{x \in \hat{X}_1 \mid \exists \xi \in C\colon \|f_i(x) - \tilde{f}(\xi)\| + \rho(x, \xi) < 2d(x)\}.$$

Each of the sets Z_i is open relatively $\hat{X}_1$, i.e., if $x_0 \in Z_i$ then there exists such an $\varepsilon > 0$ that $O^X(x_0, \varepsilon) \bigcap \hat{X}_1 \subset Z_i$. This follows from the continuity of restrictions of the maps f_i and the function d to the set $\hat{X}_1$ and the fact that $d(x_0) > 0$. Thus, each of the sets Z_i is measurable. In addition, since by construction for almost every x the set $\{f_n(x)\}$ is everywhere dense in the set $F(x)$, we get that for each of these x belonging also to the set $\hat{X}_1$, there exists at least one number i for which $x \in Z_i$. In other words, deleting from the set $\hat{X}_1$ a subset of a zero measure, we obtain $\hat{X}_1 = \bigcup_i Z_i$.

For $x \in \hat{X}_1$ set

$$\hat{f}(x) = f_i(x) \quad \forall x \in Z_i \setminus \Big(\bigcup\nolimits_{j<i} Z_j\Big).$$

Obviously, so defined on $\hat{X}_1$, the map $\hat{f}$ is measurable. Further, in the same way define the map $\hat{f}$ on the set $\hat{X}_2 \setminus \hat{X}_1$ and so on. It is clear that so defined on $X \setminus C$ the map $\hat{f}$ is a measurable selection of $\hat{F}$.

Set

$$f(x) = \begin{cases} \tilde{f}(x), & x \in C, \\ \hat{f}(x), & x \notin C. \end{cases}$$

Evidently, f is a measurable selection of the set-valued map F. Prove that f is continuous at each point $x \in C$. In fact, let $x_0 \in C$, $\{x_i\} \to x_0$ as $i \to \infty$. Let us show that $f(x_i) \to f(x_0)$. In this connection we will assume that $x_i \notin C$ for all i (since for each sequence lying in C the desired assertion follows from the continuity of $\tilde{f}$). The set-valued map F is lower semicontinuous at the point x_0. Thus for each number i there exists $y_i \in F(x_i)$ such that

$$y_i \to \tilde{f}(x_0) \Rightarrow d(x_i) \le \|y_i - \tilde{f}(x_0)\| + \rho(x_i, x_0) \to 0,$$

yielding $d(x_i) \to 0$, $i \to \infty$. By using this, we have

$$\begin{aligned} f(x_i) = \hat{f}(x_i) \in \hat{F}(x_i) &\Rightarrow \exists \xi_i \in C: \|f(x_i) - \tilde{f}(\xi_i)\| + \rho(x_i, \xi_i) < 2d(x_i) \\ &\Rightarrow \|f(x_i) - \tilde{f}(\xi_i)\| \to 0, \ \xi_i \to x_0 \\ &\Rightarrow \tilde{f}(\xi_i) \to \tilde{f}(x_0) = f(x_0) \\ &\Rightarrow f(x_i) \to f(x_0), \ i \to \infty. \end{aligned}$$

So, the selection f is the desired one. □

Now, let us prove Theorem 22.2. Take the closed set $C = B_0 \bigcup \mathrm{cl}\, B_1$. For $x \in X$ define

$$H(x) = \begin{cases} F(x), & x \in B_0, \\ \mathrm{cl}\,(\mathrm{conv}\, F(x)), & x \in X \setminus B_0, \end{cases}$$

Let us show that the set-valued map H is lower semicontinuous at each point $x_0 \in C$. Indeed, let $x_0 \in B_0$. Then $H(x_0) = F(x_0)$ and the lower semicontinuity of H at x_0 follows from the lower semicontinuity of F at the same point and the inclusion $F(x) \subset H(x)$ which is true for all x.

Now, let $x_0 \notin B_0$ and a sequence $\{x_n\}$ converges to x_0. Then, by the closedness of the set B_0 we can assume, without loss of generality, that $x_n \notin B_0$ $\forall n$ and hence $H(x_n) = \mathrm{cl}\,(\mathrm{conv}\, F(x_n))$ $\forall n$.

Let, firstly, $y_0 \in \mathrm{conv}\, F(x_0)$. Then there exist such $y_0^j \in F(x_0)$ and $\lambda_j > 0$, $j = \overline{1, m}$ that

$$\sum_j \lambda_j y_0^j = y_0, \quad \sum_j \lambda_j = 1.$$

Since, by assumption, F is lower semicontinuous at the point x_0, for each j there exists such a sequence $\{y_n^j\}$ that $y_n^j \in F(x_n)$ $\forall n$ and $y_n^j \to y^j$. Then, evidently, $y_n = \sum_j \lambda_j y_n^j \in H(x_n)$ $\forall n$ and $y_n \to y_0$. From the above reasoning it is easy to obtain that if $y_0 \in \mathrm{cl}\,(\mathrm{conv}\, F(x_0))$, then there exist $\tilde{y}_n \in F(x_n)$, for which $\tilde{y}_n \to y_0$, completing the proof of the lower semicontinuity of the set-valued map H (see also [17], Theorem 1.3.26). Similarly, by the Luzin theorem, from the measurability of F follows the measurability of the set-valued map H.

By the Michael theorem the restriction of the set-valued map H to the set C admits a continuous selection $\tilde{f}$. Let us apply Lemma 22.3 to the set-valued map H. By virtue of this assertion, there exists a measurable selection h of H which is continuous at each point $x \in C$ and $h|_C = \tilde{f}$.

For each $x \in X$ set $r(x) = \mathrm{dist}\,(h(x), F(x))$. From the condition of the theorem and by the definition of the set-valued map H it is clear that $H(x) = F(x)$ for all $x \in B = B_0 \cup B_1$. Hence $r(x) = 0$ $\forall x \in B$.

Define the set-valued map $\hat{F}$ by the relation

$$\hat{F}(x) = \begin{cases} h(x), & r(x) = 0, \\ y \in F(x) \colon \|y - h(x)\| < 2r(x), & r(x) > 0. \end{cases}$$

Repeating the reasoning similar to that used during the proof of Lemma 22.3 we obtain that the function r is measurable and the set-valued map $\hat{F}$ admits a measurable selection f. Let us show that it is the desirable one.

Indeed, by virtue of the above mentioned, f is a measurable selection of F. Let $x_0 \in B$ and $x_i \to x_0$. Prove that $f(x_i) \to f(x_0)$. Indeed, by construction, $h(x_0) = f(x_0)$, $h(x_i) \to h(x_0)$. In addition, from the lower semicontinuity of F at the points of the set B we have $r(x_i) \to 0 = r(x_0)$. Hence,

$$\begin{aligned}\|f(x_i) - f(x_0)\| &\le \|f(x_i) - h(x_i)\| + \|h(x_i) - h(x_0)\| \\ &\le 2r(x_i) + \|h(x_i) - h(x_0)\| \to 0.\end{aligned}$$

Thus $f(x_i) \to f(x_0)$ and so the map f is continuous at the point x_0 that completes the proof.

Remark 22.4. In Theorem 22.2 the regularity of the measure μ was used only for the opportunity to use the Luzin theorem.

Remark 22.5. The measurable choice theorem and the continuous selection theorem are directly used in the proof of Theorem 22.2. At the same time, formally they both follow from it. Namely, if $B_0 = B_1 = \emptyset$, then Theorem 22.2 is the measurable choice theorem, and if the set B_0 is empty and $B_1 = X$, then it turns into Theorem 21.1.

Remark 22.6. Let $C \subseteq B$ be a closed subset and $\tilde{f}$ a continuous selection of the restriction $F|_C$. Then the selection f arising in Theorem 22.2 may be chosen so that $f|_C = \tilde{f}$. This theorem on the extension from a closed subset can be deduced from Theorem 22.2 in the same way as Theorem 21.2 from the Michael theorem.

Notice that if we take $B_1 = \emptyset$, and the set B_0 consisting of a single point x_0, then Theorem 22.2 shows that if F is lower semicontinuous at x_0, then F admits a measurable selection which is continuous at the point x_0. The necessity to study such selections arises in applications.

Consider the following natural generalization of Theorem 22.2.

Theorem 22.7. *Let a set-valued map F be measurable and the following set be given: a closed set $B_0 \subset X$, which is either empty or zero-dimensional and a measurable set $B_1 \subseteq X$ such that the set $F(x)$ is convex for each $x \in B_1$. Suppose that F is lower semicontinuous at each point $x \in B_0 \bigcup \operatorname{cl} B_1$. Then F admits a denumerable set of measurable selections $\{f_i\}$, such that each map f_i is continuous at all points of the set $B = B_0 \bigcup B_1$, and for almost all x the set $\{f_1(x), f_2(x), \dots\}$ is everywhere dense in the set $F(x)$.*

Proof. Since the set B is measurable, by virtue of the regularity of the measure μ, for each positive integer l there exist such a closed set $C_l \subseteq B$ and an open set $G_l \supseteq B$, that $\mu(G_l \setminus C_l) < l^{-1}$. By Theorem 21.4 there exists a denumerable set of continuous selections $\{f_{l,i}\}_{i=1}^{\infty}$ of the set-valued map $F|_{C_l}$, such that for each $x \in C_l$ the set $\{f_{l,1}(x), f_{l,2}(x), \dots\}$ is everywhere dense in the set $F(x)$.

For arbitrary positive integers l, i and $x \in G_l$ set

$$F_{l,i}(x) = \begin{cases} f_{l,i}(x), & x \in C_l, \\ F(x), & x \in G_l \setminus C_l. \end{cases}$$

Therefore, for each positive integer l, i the set-valued map $F_{l,i}$ is defined on the set G_l. Let us show that it satisfies all assumptions of Theorem 22.2. To this end, it is sufficient to verify that the set-valued map $F_{l,i}$ is lower semicontinuous at each point of the set $B_0 \bigcup \bar{B}_{1,l}$. Here $\bar{B}_{1,l}$ is the closure of the set B_1 with respect to the space G_l, i.e., it consists of all points $\xi \in G_l$, for which there exists a sequence $\{\xi_n\}$ such that $\xi_n \in G_l \bigcap B_1$ $\forall n$ and $\xi_n \to \xi$.

In fact, let $x_0 \in B_0 \bigcup \bar{B}_{1,l}$, $y_0 \in F_{l,i}(x_0)$ and a sequence $\{x_n\} \subset G_l$ converges to the point x_0. We need to find such a sequence $\{y_n\}$ convergent to y_0 that $y_n \in F_{l,i}(x_n)$ $\forall n$. Passing to a subsequence if necessary, consider two cases.

The first case: $x_n \in C_l$ $\forall n$. Then $y_0 = f_{l,i}(x_0)$ and hence $y_n = f_{l,i}(x_n)$ gives the desired sequence since by construction $y_n \in F_{l,i}(x_n)$ $\forall n$ and $y_n \to y_0$ by virtue of the continuity of the map $f_{l,i}$.

Consider the second case: $x_n \in (G_l \setminus C_l)$ $\forall n$. Then, by construction $F_{l,i}(x_n) = F(x_n)$ $\forall n$. In addition, $x_0 \in B_0 \bigcup \operatorname{cl} B_1$, since, obviously, $\bar{B}_{1,l} \subset \operatorname{cl} B_1$. Thus the existence of the desired sequence $\{y_n\}$ follows from the assumption that F is lower semicontinuous on $B_0 \bigcup \operatorname{cl} B_1$.

Therefore, for the set-valued map $F_{l,i}$ defined on G_l all assumptions of Theorem 22.2 are fulfilled. By virtue of this theorem, there exists such a measurable map $f_{l,i}\colon G_l \to Y$, that $f_{l,i}(x) \in F_{l,i}(x)$ for all $x \in G_l$ and the map $f_{l,i}$ is continuous in all points of the set $B_0 \bigcup \operatorname{cl} B_1$. Further, by Theorem 19.7 on the measurable choice, there exists such a denumerable set of measurable selections $\{f_j\}$ of the map F that for almost all $x \in X$ the set $\{f_1(x), f_2(x), \dots\}$ is everywhere dense in the set $F(x)$.

For positive integers l, i, j define on the space X the maps $f_{l,i,j}$ by the relation

$$f_{l,i,j}(x) = \begin{cases} f_{l,i}(x), & x \in G_l, \\ f_j(x), & x \in X \setminus G_l. \end{cases}$$

Each of the maps $f_{l,i,j}$ is obviously measurable and, in addition, it is continuous in all points of the set $B_0 \bigcup \operatorname{cl} B_1$, which follows from the openness of the sets G_l and the above mentioned continuity properties of the maps $f_{l,i}$. Further, by construction, the set of points $f_{l,i,j}(x)$, where l, i, j take all possible natural values is everywhere dense in the set $F(x)$ and $f_{l,i,j}(x) \in F(x)$ for almost all $x \in X$. So, the set of selections $f_{l,i,j}$ is the desirable one. □

Consider one more type of selections. Let α be a real-valued function on X such that $\alpha(x) \in (0, 1]$ for all x.

Definition 22.8. A set-valued map F is called α-lower semi-Hölder at a point $x_0 \in X$ if there exist k and $\varepsilon > 0$ such that

$$\operatorname{dist}(y, F(x)) \leq k\rho(x, x_0)^{\alpha(x_0)} \quad \forall x \in O^X(x_0, \varepsilon), \ \forall y \in F(x_0).$$

Lemma 22.9. *Let C be a closed subset of X and a set-valued map F is α-lower semi-Hölder at each point of C. Suppose that $\tilde{f}$ is a selection of the restriction $F|_C$ satisfying the following condition: for each $x_0 \in C$ there exist k and $\varepsilon > 0$ such that*

$$\|\tilde{f}(x) - \tilde{f}(x_0)\| \leq k\rho(x, x_0)^{\alpha(x_0)} \quad \forall x \in O^X(x_0, \varepsilon) \bigcap C. \tag{22.2}$$

Then F has such a measurable selection f, that $\tilde{f} = f|_C$ and for each $x_0 \in C$ there exist k and $\varepsilon > 0$ such that

$$\|f(x) - f(x_0)\| \leq k\rho(x, x_0)^{\alpha(x_0)} \quad \forall x \in O^X(x_0, \varepsilon), \tag{22.3}$$

The proof of this assertion is analogous to the proof of Lemma 22.3.

Condition (22.2) differs from (22.3) so that in (22.2) points x are taken only from the set C. It is natural to call a selection satisfying condition (22.3) *α-Hölder on the set C*. From the formulated lemma we can deduce assertions analogous to Theorem 22.2 concerning the existence of measurable selections being α-Hölder on the set C. In this case it is appropriate to use, instead of the Michael theorem, sufficient conditions for the existence of α-Hölder selections for the restriction $F|_C$. For example, it is possible to apply Theorem 21.6 on the existence of a Lipschitz selection.

23 Differential inclusions

Nonstrictly speaking, a differential inclusion is a differential equation with a set-valued right-hand side.[1] In a sufficiently general form, the differential inclusion can be written as

$$x'(t) \in F(t, x(t)), \tag{23.1}$$

where $t \in \mathbb{R}$ is the time parameter, $x(t)$ is a function defined on a certain subset of $\mathbb{R}$ with values in $\mathbb{R}^n$, and $F = F(t, x)$ is a set-valued map that associates to every $t \in \mathbb{R}$ and $x \in \mathbb{R}^n$ a nonempty closed subset $F(t, x) \subset \mathbb{R}^n$.

The theory of differential inclusions arose in the 1930s, but its active development, stimulated by numerous applications in control theory, optimization theory, theory of differential equations with discontinuous right-hand sides, mathematical economics and other fields, started essentially later, in the 1950s and 1960s.

Let us dwell on problems of control theory and theory of differential equations with discontinuous right-hand sides. A sufficiently adequate mathematical model of a feedback control system may be described by relations

$$\begin{cases} x'(t) = f(t, x(t), u(t)) \\ u(t) \in U(t, x(t)). \end{cases}$$

Here, the function f characterizes the dynamics of the system, $u(\cdot)$ is the control function, and $x(\cdot)$ is the corresponding trajectory. The set-valued map U describes restrictions on the control, it is often called *the feedback set-valued map*. Along with this control system, consider the differential inclusion

$$x'(t) \in F(t, x(t)) = f(t, x(t), U(t, x(t))). \tag{23.2}$$

It is clear that if a pair of functions $(x(\cdot), u(\cdot))$ presents a trajectory of the control system and a corresponding control, then this trajectory is the solution of differential inclusion (23.2). However, the inverse passage from inclusion (23.2) to the control system is not so evident. The equivalence of the control system to differential inclusion (23.2) is established with the help of the Filippov theorem (see Theorem 19.9).

A further important application for differential inclusions that was found around the same time was their use for the study of differential equations with discontinuous right-hand sides. Namely, in the classical theory of differential equations the right side of the equations $f(t, x)$ usually is assumed to be continuous in the variable x. However, some problems of mechanics and engineering, theory of differential games, etc. lead to differential equations with discontinuous right-hand sides. However, for such differential equations is not quite clear what to call its solution. Suppose, for example,

1 The exposition here partly follows [17] and [62], Chapter 2.

DOI 10.1515/9783110460308-024

$f(x) = -\operatorname{sgn} x = \{1, \text{at } x < 0,\ -1, \text{at } x \geq 0\}$. Then the Cauchy problem

$$x'(t) = f(x(t)), \quad x(0) = 0,$$

as it is not difficult to see, has no solution in the ordinary sense (verify it as an exercise).

Thus, for differential equations with discontinuous right-hand sides there is a need to determine a solution in some new way, but this must be done correctly in a mathematical and in a physical sense. In this case the following approach turned out to be the most natural (see [62]).

Let a function $f(t, x)$ be bounded in a certain neighborhood of each point and the set M of its points of discontinuity has zero measure. For a fixed (t, x) by $F(t, x)$ denote the convex closure of all limit points of $f(\tilde{t}, \tilde{x})$, when $(\tilde{t}, \tilde{x}) \notin M$ and $(\tilde{t}, \tilde{x}) \to (t, x)$. By a solution of the initial differential equation $x'(t) = f(t, x(t))$ we will mean a solution of the corresponding differential inclusion (23.1). It turns out that this approach to the original differential equation, essentially replacing it with the corresponding differential inclusion, is very effective (see [62]).

By the Cauchy problem for differential inclusion (23.1) we mean

$$x'(t) \in F(t, x(t)), \quad x(t_0) = x_0. \tag{23.3}$$

It consists in finding an absolutely continuous function[2] $x(\cdot)$, which for some $\bar{t} > t_0$ satisfies the above inclusion for almost all $t \in (t_0, \bar{t})$ and the initial condition $x(t_0) = x_0$ for given t_0 and x_0. The function $x(\cdot)$ is called its *local solution.*

The Cauchy problem for differential inclusions has much in common with the Cauchy problem

$$x'(t) = f(t, x(t)), \quad x(t_0) = x_0 \tag{23.4}$$

for classical ordinary differential equations. Namely, under very general assumptions, mainly lying in the fact that the right-hand side f is measurable in time t and continuous in the phase variable x, the Cauchy problem (23.4) has a local solution. Similarly, as it will be shown below, the Cauchy problem for a differential inclusion also has a local solution under very general assumptions on the set-valued map F.

At the same time, there are fundamental differences between differential equations and differential inclusions. For example, we know that if the right-hand side f satisfies the Lipschitz condition with respect to x in a certain neighborhood of (t_0, x_0) uniformly with respect to t, then the local solution of Cauchy problem (23.4) is unique. For the differential inclusions the question of uniqueness of the solution of the Cauchy

2 Recall that a function $x(\cdot)$ is called absolutely continuous on an interval if it is differentiable in a.e. points of it, its derivative is integrable and $x(\cdot)$ may be presented in the form of the integral with variable upper limit of its derivative. Main properties of absolutely continuous functions may be found, e.g., in [27].

problem is quite different. There are no reasonable assumptions about the set-valued map F (naturally, except that for every (t, x) the set $F(t, x)$ consists of a single point), under which the solution of Cauchy problem (23.3) is unique. The simplest example is the Cauchy problem with constant right-hand part

$$x'(t) \in A, \quad x(t_0) = x_0,$$

where A is a given subset of $\mathbb{R}^n$, consisting of more than one point. Obviously, this problem has an infinite number of solutions.

Let us formulate and prove the theorem on the existence of a local solution to the Cauchy problem for differential inclusions.

Theorem 23.1. *Let G be an open subset of $\mathbb{R} \times \mathbb{R}^n$ on which an upper semicontinuous set-valued map F with compact convex values in $\mathbb{R}^n$ is given. Then for each point $(t_0, x_0) \in G$ there exists a local solution to Cauchy problem* (23.3). *Moreover, for each $\alpha, \beta > 0$ such that*

$$Z = [t_0, t_0 + \alpha] \times O(x_0, \beta) \subset G,$$

for

$$d = \min\{\alpha, \beta/m\}, \quad \textit{where } m = \sup\{|y|,\ y \in F(t, x),\ (t, x) \in Z\}$$

this solution is defined on the interval $[t_0, t_0 + d]$.

Note that from Proposition 17.30 it follows that the least upper bound m is a finite number. The proof of this theorem (we do it following [62]) will be preceded by some statements, some of which are known, but for the sake of completeness, we prove them here as well.

Proposition 23.2. *Let D be a convex compact subset of $\mathbb{R}^n$, $v\colon [a, b] \to D$ a measurable function. Then*

$$(b-a)^{-1} \int_a^b v(t)\,dt \in D.$$

Proof. Set

$$d = (b-a)^{-1} \int_a^b v(t)\,dt$$

and suppose that $d \notin D$. Then by the theorem on the strict separability of convex sets (see Theorem 4.17 in Part I) there exist such $l \in \mathbb{R}^n$ and a number α, that $\langle l, x\rangle \le \alpha\ \forall x \in D$ and $\langle l, d\rangle \gg \alpha$. Hence we have

$$\alpha < \langle l, d\rangle = (b-a)^{-1} \int_a^b \langle l, v(t)\rangle\,dt \le (b-a)^{-1} \int_a^b \alpha\,dt = \alpha.$$

This contradiction completes the proof. □

Proposition 23.3. *Let D be a convex compact set in $\mathbb{R}^n$ and $\{x_k\}$ a sequence of absolutely continuous functions $x_k\colon [a, b] \to \mathbb{R}^n$ such that $x_k(t) \to x(t)$ for every t. Suppose also that for each k for almost all $t \in [a, b]$ we have $x'_k(t) \in D$. Then the function $x(\cdot)$ is also absolutely continuous and, moreover, $x'(t) \in D$ for almost all $t \in [a, b]$.*

Proof. Since the set D is bounded, there exists a $c > 0$ such that $|x_k(t)| \le c$ for all k and almost all $t \in [a, b]$. Take arbitrary $t_1, t_2 \in [a, b]$. By using the Newton–Leibnitz formula, we have

$$|x_k(t_2) - x_k(t_1)| = \left| \int_{t_1}^{t_2} x'_k(t)\,dt \right| \le \int_{t_1}^{t_2} |x'_k(t)|\,dt \le c\,|t_2 - t_1|.$$

Passing in this inequality to the limit as $k \to \infty$, we get

$$|x(t_2) - x(t_1)| \le c|t_2 - t_1| \quad \forall t_1, t_2 \in [a, b].$$

Therefore, the function $x(\cdot)$ satisfies the Lipschitz condition and hence it is absolutely continuous (see, e.g., [27]).

By Proposition 23.2 for almost all $t \in (a, b)$ and for a sufficiently small $\tau > 0$ we have

$$\tau^{-1}(x_k(t + \tau) - x_k(\tau)) = \tau^{-1} \int_t^{t+\tau} x'_k(t)\,dt \in D.$$

Passing to the limit as $k \to \infty$, we get

$$\tau^{-1}(x(t + \tau) - x(\tau)) \in D. \tag{23.5}$$

But the function $x(\cdot)$ is absolutely continuous and hence it is almost everywhere differentiable. Therefore, from (23.5) as $\tau \to 0$ we obtain $x'(t) \in D$ for almost all $t \in [a, b]$. □

Below, for convenience, for $\delta > 0$ by M^δ we will denote the closed δ-neighborhood of a set M, i.e., $M^\delta = B(M, \delta)$. For $\delta > 0$ an absolutely continuous function $y\colon [a, b] \to \mathbb{R}^n$ is called *a δ-solution (or an approximate solution, up to δ) of differential inclusion* (23.1) *on the interval* $[a, b]$, if

$$y'(t) \in F_\delta(t, y(t)) \quad \text{for almost all } t \in [a, b],$$

where

$$F_\delta(t, y) = (\operatorname{conv} F(t^\delta, y^\delta))^\delta.$$

Lemma 23.4. *Suppose that F satisfies all assumptions of Theorem 23.1. Let $\delta_k \to +0$, functions x_k are δ_k-solutions of differential inclusion* (23.1) *on an interval $[a, b]$ and the sequence of functions $\{x_k\}$ uniformly converges to a function $x(\cdot)$, and furthermore $(t, x(t)) \in G\ \forall t \in [a, b]$. Then the limit function $x(\cdot)$ is the solution of* (23.1) *on the interval* $[a, b]$.

Proof. The function $x(\cdot)$ is continuous as the uniform limit of continuous functions. Fix an arbitrary $\tau \in [a, b]$ and take any $\varepsilon > 0$. From the condition that F is upper semicontinuous it follows that there exists an $\eta > 0$ such that

$$F(t, x) \subset A^{\varepsilon}, \quad \forall (t, x) \in G_0, \tag{23.6}$$

where

$$G_0 = \{(t, x) \colon |t - \tau| \le 2\eta,\ |x - x(\tau)| \le 3\eta\}, \quad A = F(\tau, x(\tau)).$$

By the continuity of the function $x(\cdot)$, there exist such a $\gamma \in (0, \eta)$ and an integer k_0, that

$$\delta_k < \min\{\eta, \varepsilon\}, \quad |x_k(t) - x(t)| \le \eta, \quad |x(t) - x(\tau)| \le \eta$$

for all t such that $|t - \tau| < \gamma$ and $k > k_0$. From here and (23.6) for $\delta = \delta_k$, $k > k_0$, $|t - \tau| < \gamma < \eta$, we have

$$t^{\delta} \subset \tau^{2\eta}, \quad (x_k(t))^{\delta} \subset (x(\tau))^{3\eta} \quad \to F(t^{\delta}, (x_k(t))^{\delta}) \subset A^{\varepsilon}.$$

Since the function x_k is a δ_k-solution, and the set A is convex, for almost all t we have

$$x_k'(t) \in (\operatorname{conv} F(t^{\delta}, (x_k(t))^{\delta})^{\delta} \subset (\operatorname{conv} A^{\varepsilon})^{\delta} \subset A^{2\varepsilon}.$$

Therefore, by Proposition 23.3 the function $x(\cdot)$ is absolutely continuous on the set $\{t \colon |t - \tau| < \gamma\}$ and $x'(t) \in A^{2\varepsilon}$ for almost all t in this set.

Sets of the type $\{t \colon |t - \tau| < \gamma\}$ form an open covering of the interval $[a, b]$. Choosing a finite subcovering, we obtain that the function $x(\cdot)$ is absolutely continuous on $[a, b]$. In addition, it is proved that for every $\tau \in (a, b)$ at which the function $x(\cdot)$ is differentiable, we have $x'(\tau) \in F(\tau, x(\tau))^{2\varepsilon}$ for an arbitrary $\varepsilon > 0$. Passing in this inclusion to the limit as $\varepsilon \to 0$, due to the closedness of the set $F(\tau, x(\tau))$ we get $x'(\tau) \in F(\tau, x(\tau)$ which shows that $x(\cdot)$ is a solution of (23.1). □

Let us return to the proof of Theorem 23.1 in which we will use the method of Euler's polygonal lines. For $k = 1, 2, \ldots$ set

$$h_k = d/k, \quad t_{k,i} = t_0 + i h_k, \quad i = 0, 1, \ldots, k.$$

Let us construct a polygonal line $x_k(t)$. Put $x_k(t_{k,0}) = x_0$. If for a certain i the value $x_k(t_{k,i})$ is already defined and

$$|x_k(t_{k,i}) - x_0| \le m|t_{k,i} - t_0|, \tag{23.7}$$

then for $t \in (t_{k,i}, t_{k,i+1}]$ we define $x_k(t)$ by the formula

$$x_k(t) = x_k(t_{k,i}) + (t - t_{k,i}) v_{k,i}, \tag{23.8}$$

taking any $v_{k,i} \in F(t_{k,i}, x_k(t_{k,i}))$. Since by (23.7) $(t_{k,i}, x_k(t_{k,i})) \in Z$, then $|v_{k,i}| \le m$, and from (23.7) and (23.8) we have

$$|x_k(t) - x_0| \le |x_k(t) - x_k(t_{k,i})| + |x_k(t_{k,i}) - x_0| \le m|t - t_0| \quad \forall t \in (t_{k,i}, t_{k,i+1}]. \tag{23.9}$$

At the same time the inequality obtained from (23.7) by the replacing of i with $i+1$ also holds true.

Thus, a polygonal line x_k can be built step by step on intervals $[t_{k,i}, t_{k,i+1}]$ for $i = 0, 1, \dots, k-1$. By (23.9) we have $(t, x_k(t)) \in Z$ $\forall t \in [t_0, t_0 + d]$. From (23.8) it follows that the function x_k is absolutely continuous and $|x_k'(t)| \le m$ $\forall t \ne t_{k,i}$. Since

$$x_k'(t) = v_{k,i} \in F(t_{k,i}, x_k(t_{k,i}))$$

and

$$|x_k(t) - x_k(t_{k,i})| \le mh_k \quad \forall t \in [t_{k,i}, t_{k,i} + h_k],$$

we get that x_k is a δ_k-solution of inclusion (23.3) for $\delta_k = h_k \max\{1, m\}$ with $\delta_k \to 0$, $k \to \infty$.

By virtue of (23.9) the sequence of functions $\{x_k\}$ is uniformly bounded, and since $|x_k'(t)| \le m$ it is equicontinuous. By the Arzela[3]–Ascoli[4] theorem (see [27]) a uniformly convergent subsequence can be selected from this sequence. By Lemma 23.4 its limit $x(\cdot)$ is a solution of (23.1). From the fact that $x_k(t_0) = x_0$ for all k, we have $x(t_0) = x_0$ concluding the proof.

In the last section of this chapter we will obtain local and global existence theorems for differential inclusions by the fixed point method. More details about differential inclusions theory may be found, for example, in [15, 63, 17, 5, 19, 62, 20, 22].

3 Cesare Arzelà (1847–1912), an Italian mathematician.
4 Giulio Ascoli (1843–1896), an Italian mathematician.

24 Fixed points and coincidences of maps in metric spaces

24.1 The case of single-valued maps

For a certain set X, let a map $f: A \subseteq X \to X$ be given. A point $x_0 \in A$ is called *a fixed point of the map* f if $x_0 = f(x_0)$. *Fixed point theorems* are various assertions containing sufficient conditions posed on sets A, X and a map f which guarantee the existence of at least one fixed point of f. Corresponding results form *the fixed point theory* which is of great importance both for theoretical and applied aspects of contemporary mathematics. A wide variety of existence problems for many types of operator and differential equations and inclusions, problems of functional analysis, topology, theory of control systems, dynamical systems, theory of games, mathematical economics and other branches may be reduced to the existence of fixed points of appropriate maps. Some such applications will be demonstrated below.

Let us present here the best known fixed point result for a class of maps in a metric space which has found numerous applications. It is the Banach fixed point principle for contraction maps which may be formulated in the following way.

Theorem 24.1. *Let (X, ρ_X) be a complete metric space and $f: X \to X$ a contraction map, i.e., there exists such a $k < 1$ that*

$$\rho_X(f(x_1), f(x_2)) \le k\rho_X(x_1, x_2) \quad \forall x_1, x_2 \in X.$$

Then the map f has a fixed point x_0 and, moreover, it is unique and for an arbitrary point $\xi \in X$ the sequence of iterations $f^{(n)}(\xi)$ converges to x_0 while $n \to \infty$.

We do not give the proof of this theorem here, although it is fairly simple since it follows from a more general assertion which we will obtain below.

A natural development of the notion of a fixed point is the concept of *a coincidence point for a pair of maps.*

Let X and Y be arbitrary sets and $\psi, \varphi: X \to Y$ given maps.

Definition 24.2. A point $x_0 \in X$ is a coincidence point of maps ψ and φ if $\psi(x_0) = \varphi(x_0)$.

It is clear that if $X = Y$ and $\psi(x) \equiv x$ is the identity map, a coincidence point of ψ and φ turns out to be a fixed point of φ.

In the following we will assume that (X, ρ_X) and (Y, ρ_Y) are metric spaces. The problem is to find conditions on maps ψ and ϕ which guarantee the existence of their coincidence point. In this and in the next sections we will use the results from [64, 65].

DOI 10.1515/9783110460308-025

Definition 24.3. For a given $\alpha > 0$, a map $\psi\colon X \to Y$ is called α-covering if

$$\psi(B^X(x,r)) \supseteq B^Y(\psi(x), \alpha r) \quad \forall r \geq 0\ \forall x \in X. \tag{24.1}$$

A map is said to be covering if it is α-covering for a certain $\alpha > 0$.

Consider the notion of a covering map. It is easy to see that a covering map is surjective. It is also evident that if for a certain $\alpha > 0$ a map ψ is α-covering then it is γ-covering also for arbitrary $\gamma \leq \alpha$. The simplest example of a 1-covering map $\psi\colon X \to X$ is the identity map $\Psi(x) \equiv x$.

Notice that a covering map can be discontinuous. As an example, take the space $X = [0,1]$ with the natural metric, the two-point space $Y = \{0,1\}$, and the Dirichlet[1] function $\psi\colon X \to Y$ assigning $y = 1$ to each rational x and $y = 0$ to each irrational x. This map is obviously α-covering for each $\alpha > 0$ but is discontinuous at each point.

Now, let X and Y be Banach spaces. A bounded linear operator $A\colon X \to Y$ is α-covering for a certain $\alpha > 0$ if and only if its image $\operatorname{im} A$ coincides with the whole space Y. This assertion makes up the Banach open map theorem (see, e.g., [27]).

More generally, let a map ψ be Fréchet[2] continuously differentiable on the whole space X and there exists such a $\delta > 0$ that $\psi'(x)(B^X(0,1)) \supset B^Y(0,\delta)$ $\forall x \in X$. Then the map ψ is covering.

The situation in metric spaces is essentially more complicated. Let, for example, $X = Y = [0,1]$ with the natural metric and $\psi(x) \equiv x$ be the identity map. As we mentioned already, it is 1-covering. At the same time, any perturbation close to it of the form $\psi_\varepsilon(x) \equiv (1-\varepsilon)x + \varepsilon$ for each small $\varepsilon > 0$ is not covering since it is not surjective. Thus, the covering property of maps in metric spaces is not stable with respect to small perturbations.

Now, let us clear up how the α-covering property behaves under limit passages. Namely, for a given $\alpha > 0$, take a sequence of α-covering maps $\{\psi_n\}$ and a map ψ. Suppose that the pointwise convergence holds true:

$$\rho_Y(\psi_n(x), \psi(x)) \to 0, \ n \to \infty \quad \forall x \in X.$$

We can ask, should the map ψ be covering? The next example gives a negative answer to this question (even if the spaces X and Y are complete).

Example 24.4. Let $X = Y = [0,2]$, and each of functions ψ_n is piecewise linear: for $x \in [0,1/n]$ its graph linearly connects the point $(0,0)$ with the point $M_n = (1/n, 1)$, and for $x \in [1/n, 2]$ it linearly connects the point M_n with the point $(2,2)$. The function ψ is defined from the conditions

$$\psi(x) = \begin{cases} 0, & x = 0, \\ 1 + x/2, & x \in (0,2]. \end{cases}$$

1 Johann Peter Gustav Lejeune Dirichlet (1805–1859), a German mathematician.
2 Maurice René Fréchet (1878–1973), a French mathematician.

It is obvious that $\psi_n(x) \to \psi(x)$, $n \to \infty$ $\forall x$, each of maps ψ_n is $1/2$-covering, however the map ψ is not covering (since $1 \notin \psi(X)$).

At the same time, in the next section we will prove that under uniform convergence the covering property is preserved, i.e., if a sequence of α-covering maps converges then its limit is a covering map.

Let us formulate and prove a theorem on the existence of coincidence points.

Definition 24.5. For $\beta \geq 0$, we will say that a map $\varphi\colon X \to Y$ is β-Lipschitz if it satisfies the Lipschitz condition with the constant β, i.e.,

$$\rho_Y(\varphi(x_1), \varphi(x_2)) \leq \beta \rho_X(x_1, x_2) \quad \forall x_1, x_2 \in X.$$

Thus, if $X = Y$, the map φ is contracting provided it is β-Lipschitz for a certain $\beta < 1$.

Theorem 24.6. *Let $\alpha > \beta$ be given. Suppose that the space X is complete, a map ψ is α-covering and closed (i.e., its graph is closed) and a map φ is β-Lipschitz. Then for an arbitrary $x \in X$ there exists such a $\xi = \xi(x) \in X$, that*

$$\psi(\xi) = \varphi(\xi), \quad \rho_X(\xi, x) \leq \frac{\rho_Y(\psi(x), \varphi(x))}{\alpha - \beta}. \tag{24.2}$$

Proof. Dividing the metric ρ_Y over α, we will assume, without loss of generality, that $\alpha = 1$ and hence $\beta < 1$. The proof of the theorem is based on the iteration method. Fix arbitrary $x \in X$ and set $x_0 = x$. By the covering condition (24.1) there exists $x_1 \in X$ such that

$$\psi(x_1) = \varphi(x_0), \quad \rho_X(x_0, x_1) \leq \rho_Y(\psi(x_0), \varphi(x_0)). \tag{24.3}$$

Construct by induction such a sequence $\{x_i\}$, that for $i \geq 2$ we have

$$\psi(x_i) = \varphi(x_{i-1}), \quad \rho_X(x_i, x_{i-1}) \leq \beta \rho_X(x_{i-1}, x_{i-2}). \tag{24.4}$$

Indeed, let the desired $x_0, x_1, \dots, x_i$ be constructed already. Then by virtue of (24.1) there exists $x_{i+1} \in X$ such that

$$\psi(x_{i+1}) = \varphi(x_i), \quad \rho_X(x_{i+1}, x_i) \leq \rho_Y(\psi(x_i), \varphi(x_i)).$$

But then

$$\rho_X(x_{i+1}, x_i) \leq \rho_Y(\psi(x_i), \varphi(x_i)) = \rho_Y(\varphi(x_{i-1}), \varphi(x_i)) \leq \beta \rho_X(x_i, x_{i-1}),$$

where the last inequality follows from the Lipschitz condition for φ. The construction of x_{i+1}, and the whole desired sequence is complete.

Show that $\{x_i\}$ is the Cauchy sequence. Indeed, take arbitrary numbers $j > l$. By the triangle inequality we have

$$\rho_X(x_j, x_l) \leq \rho_X(x_j, x_{j-1}) + \cdots + \rho_X(x_{l+1}, x_l).$$

But by virtue of inequality (24.4) we have

$$\rho_X(x_{s+1}, x_s) \le \beta^s \rho_X(x_1, x_0) \quad \forall s.$$

By using the sum of the geometric progression formula we get

$$\rho_X(x_j, x_l) \le \sum_{s=l}^{j-1} \beta^s \rho_X(x_1, x_0) \le \beta^l (1-\beta)^{-1} \rho_X(x_1, x_0)$$

and therefore $\rho_X(x_j, x_l) \to 0$ as $j, l \to \infty$ yielding that $\{x_i\}$ is the Cauchy sequence. Moreover, by virtue of (24.3) we get

$$\rho_X(x_i, x_0) \le (1-\beta)^{-1} \rho_Y(\psi(x_0), \varphi(x_0)) \quad \forall i. \tag{24.5}$$

Since the space X is complete, the sequence $\{x_i\}$ converges to a certain point ξ. By the continuity of φ, the sequence $\{\varphi(x_i)\}$ also converges. So, by the equality from (24.4) the sequence $\{\psi(x_i)\}$ also converges, and, moreover, by the closedness of ψ it converges to $\psi(\xi)$. By the mentioned equality, from (24.4) we have $\psi(\xi) = \varphi(\xi)$. Passing in (24.5) to the limit as $i \to \infty$ we finally get (24.2). □

Consider the proved theorem. First, if we take $X = Y$, and $\psi(x) \equiv x$ we get the existence of a fixed point for a contraction map (Theorem 24.1).

Further, Theorem 24.6 yields the Milyutin[3] perturbation theorem which he obtained in the 1970s. It can be formulated in the following way. Let Y be a normed space and $\psi : X \to Y$ an α-covering closed map. Assume that a map $\varphi : X \to Y$ satisfies the Lipschitz condition with a constant $\beta < \alpha$. The perturbation theorem states that then (the "perturbed") map $(\psi + \varphi)$ is $(\alpha - \beta)$-covering.

The fundamental difference between Theorem 24.6 and the contraction map principle consists in the fact that under the conditions of Theorem 24.6 a coincidence point may not be unique. As an example, we can consider an arbitrary surjective linear operator between Banach spaces with a nonzero kernel. In this connection, the following question arises.

Let all assumptions of Theorem 24.6 hold true. Then the set of coincidence points ξ for which (24.2) is valid defines the set-valued map of a variable $x \in X$. The question is: does this map admit a continuous selection $\xi(\cdot)$? The negative answer gives the following example.

Example 24.7. Let $X = [-1, 1]$ and $Y = [0, 1]$ be intervals with the natural metrics. The maps ψ and φ are defined by the relations $\psi(x) \equiv |x|$, $\varphi(x) \equiv 1/2$. Then ψ is 1-covering and the equation $|x| = 1/2$ has exactly two solutions: $x_1 = -1/2$ and $x_2 = 1/2$. Thus, the function $\xi(\cdot)$ takes no more than two values and, moreover, by (24.2) we get $\xi(-1/2) = -1/2$, $\xi(1/2) = 1/2$ and hence the function $\xi(\cdot)$ is discontinuous at zero.

3 Aleksei Alekseevich Milyutin (1925–2001), a Russian mathematician.

For applications it is important that under some natural assumptions on the maps ψ and φ their coincidence point should be stable with respect to small perturbations of these maps. Namely, let the space X be complete and a point x_0 be a coincidence point of given maps ψ, φ with $\alpha > \beta \geq 0$.

Further, let the sequences of maps $\{\psi_n\}$, $\{\varphi_n\}$ be such that for each n a map ψ_n is α-covering and closed and a map φ_n is β-Lipschitz. Suppose also that

$$\rho_Y(\psi_n(x_0), \psi(x_0)) \to 0, \quad \rho_Y(\varphi_n(x_0), \varphi(x_0)) \to 0, \quad n \to \infty.$$

Then the maps $\{\psi_n\}$, $\{\varphi_n\}$ possess such coincidence points x_n, that $x_n \to x_0$ as $n \to \infty$. In the following we will prove this assertion in an essentially more general form and for set-valued maps.

In conclusion, let us consider the following question. During the definition of an α-covering map we used closed balls in inclusion (24.1). In this definition we may replace closed balls with open ones, considering maps satisfying the condition

$$\psi(O^X(x, r)) \supseteq O^Y(\psi(x), \alpha r) \quad \forall r \geq 0, \ \forall x \in X. \tag{24.6}$$

For determinacy, such maps will be called openly α-covering (in the sense of (24.6)). We can ask, how are the notions of α-covering and openly α-covering maps connected? From the definitions it immediately follows that if a map is α-covering then it is openly α-covering and if a map is openly α-covering, then it is $(\alpha - \varepsilon)$-covering for an arbitrary $0 < \varepsilon < \alpha$. At the same time, an openly α-covering map cannot be α-covering, as the following example demonstrates.

Example 24.8. Let $X = l_1$, $Y = \mathbb{R}$. Define on the space l_1 the continuous linear functional

$$x^* = (0, 1/2, 2/3, \ldots, i/(i+1), \ldots) \in l_\infty, \tag{24.7}$$

and for $x \in l_1$ set $\psi(x) = \langle x^*, x\rangle$. Then the map ψ is $(1-\varepsilon)$-covering for each $0 < \varepsilon < 1$, since $\langle x^*, e_i\rangle \to 1 - 0$ as $i \to \infty$, where $e_i \in l_1$ is the element having 1 on the i-th position and zeros on other ones. So, the map ψ is openly 1-covering, however it is not 1-covering since if $\|x\|_{l_1} \leq 1$, then, by construction $|\langle x^*, x\rangle| < 1$.

From the above mentioned follows the next version of Theorem 24.6 which is valid for openly α-covering maps.

Proposition 24.9. *Let the space X be complete, the map ψ be openly α-covering and the map φ satisfy the Lipschitz condition with a constant $\beta < \alpha$.*

Then for arbitrary $x \in X$ and $\varepsilon > 0$ there exists such a $\xi \in X$, that

$$\psi(\xi) = \varphi(\xi), \quad \rho_X(\xi, x) \leq \frac{\rho_Y(\psi(x), \varphi(x))}{\alpha - \beta} + \varepsilon. \tag{24.8}$$

A map ψ from Example 24.8 and a constant map $\varphi(x) \equiv 1$ show (for $x = 0$) that, generally speaking, it is impossible to take $\varepsilon = 0$ in estimate (24.8).

24.2 The case of set-valued maps

Let us transfer the results of the previous section to the case of set-valued maps Ψ, Φ and, in particular, obtain the set-valued analogs of the coincidence result.

As earlier, let (X, ρ_X) and (Y, ρ_Y) be metric spaces. Suppose that Ψ and Φ are set-valued maps from X to Y with nonempty closed values.

Definition 24.10. A point $x_0 \in X$ is called a coincidence point of set-valued maps Ψ and Φ provided $\Psi(x) \bigcap \Phi(x) \neq \emptyset$.

The definition of an α-covering set-valued map formally is the same as for a single-valued one.

Definition 24.11. For a given $\alpha > 0$, a set-valued map Ψ is called α-covering if

$$\Psi(B^X(x, r)) \supseteq B^Y(\Psi(x), \alpha r) \quad \forall r \geq 0\ \forall x \in X. \tag{24.9}$$

A set-valued map is said to be covering if it is α-covering for a certain $\alpha > 0$.

Definition 24.12. A set-valued map Φ is called β-Lipschitz if it satisfies the following relation with respect to the Hausdorff metric h on the space of closed subsets of Y:

$$h(\Phi(x_1), \Phi(x_2)) \leq \beta \rho_X(x_1, x_2) \quad \forall x_1, x_2 \in X.$$

As mentioned earlier, in the case of unbounded sets the Hausdorff metric can take the infinite value (in this case the Hausdorff metric is called generalized).

Our ensuing reasoning will be based on the following set-theoretic lemma. For $x \in X$ and $r \geq 0$ set

$$N(x, r) = \Phi(x) \bigcap B^Y(\Psi(x), r).$$

It is obvious that for all $r > \operatorname{dist}(\Psi(x), \Phi(x))$ the sets $N(x, r)$ are nonempty.

Lemma 24.13. *Let a set-valued map Ψ be 1-covering and Φ be β-Lipschitz.*

Then for arbitrary $\delta > 0$, $x_0 \in X$, $r_0 > \operatorname{dist}(\Psi(x_0), \Phi(x_0))$ and $y_1 \in N(x_0, r_0)$ there exist such sequences $\{x_i\}$ and $\{y_i\}$, that

$$\rho_X(x_0, x_1) \leq r_0, \tag{24.10}$$

$$y_i \in \Psi(x_i) \bigcap \Phi(x_{i-1}), \quad \forall i \geq 1, \tag{24.11}$$

$$\rho_X(x_i, x_{i-1}) \leq (\beta + \delta) \rho_X(x_{i-1}, x_{i-2}), \ \forall i \geq 2, \tag{24.12}$$

$$\rho_Y(y_i, y_{i-1}) \leq (\beta + \delta) \rho_X(x_{i-1}, x_{i-2}), \ \forall i \geq 2. \tag{24.13}$$

Proof. By condition $y_1 \in N(x_0, r_0)$. Hence by virtue of (24.1) there exists an $x_1 \in B^X(x_0, r_0)$ such that $y_1 \in \Psi(x_1)$ and hence $y_1 \in \Psi(x_1) \bigcap \Phi(x_0)$. The desired x_1 is constructed.

If $x_0 = x_1$, then for all i set $x_i = x_1$, $y_i = y_1$. Let $x_0 \neq x_1$. The further construction will be carried out by induction. First let us construct x_2, y_2. Set $r_1 = (\beta + \delta)\rho_X(x_1, x_0)$. By the Lipschitz condition for Φ we have $h(\Phi(x_0), \Phi(x_1)) < r_1$ and hence $B^Y(\Phi(x_1), r_1) \supseteq \Phi(x_0) \ni y_1$. Thus, there exists $y_2 \in \Phi(x_1)$ such that $\rho_Y(y_1, y_2) \leq r_1$, from which, since $y_1 \in \Psi(x_1)$, it follows that $y_2 \in B^Y(\Psi(x_1), r_1)$ and hence, by virtue of (24.1) we get

$$y_2 \in \Psi(B^X(x_1, r_1)) \Rightarrow \exists x_2 : \rho_X(x_1, x_2) \leq r_1, \quad y_2 \in \Psi(x_2).$$

The desired x_2, y_2 are constructed.

Let the points $x_1, y_1, \dots, x_j, y_j$ satisfying (24.11)–(24.13) be constructed already. Construct x_{j+1}, y_{j+1}. If $x_j = x_{j-1}$, set $x_{j+1} = x_j$, $y_{j+1} = y_j$. Let $x_j \neq x_{j-1}$. Set $r_j = (\beta + \delta)\rho_X(x_j, x_{j-1})$. From the Lipschitz condition for Φ and (24.1) we have

$$\begin{aligned} h(\Phi(x_{j-1}), \Phi(x_j)) < r_j &\Rightarrow B^Y(\Phi(x_j), r_j) \supseteq \Phi(x_{j-1}) \ni y_j \\ &\Rightarrow \exists y_{j+1} \in \Phi(x_j) : \rho_Y(y_{j+1}, y_j) \leq r_j, \end{aligned}$$

from which, by virtue of (24.11) we get $y_{j+1} \in B^Y(\Psi(x_j), r_j)$ and hence by (24.1) there exists x_{j+1} such that

$$y_{j+1} \in \Psi(x_{j+1}), \quad \rho_X(x_{j+1}, x_j) \leq r_j.$$

Thus, $y_{j+1} \in \Psi(x_{j+1}) \bigcap \Phi(x_j)$ and hence for constructed x_{j+1}, y_{j+1} conditions (24.11)–(24.13) are satisfied. □

Remark 24.14. From the proof of the lemma it follows that if for all $x_1, x_2 \in X$, $y \in Y$ the infimums in the expressions dist $(\Psi(x_1), \Phi(x_2))$, dist $(y, \Psi(x_1))$ and $h(\Phi(x_1), \Phi(x_2))$ are achieved (for example, if for all x both values $\Psi(x)$ and $\Phi(x)$ are compact), then in the statement of the lemma we can take $\delta = 0$ and $r_0 = \operatorname{dist}(\Psi(x_0), \Phi(x_0))$.

Theorem 24.15. *Let $\alpha > \beta \geq 0$ and $\varepsilon > 0$ be given. Suppose that a set-valued map Ψ is α-covering and closed and a set-valued map Φ is β-Lipschitz. Let also at least one of the graphs* gph (Ψ) *or* gph (Φ) *be a complete space.*[4]

Then for arbitrary

$$x \in X, \quad r > \operatorname{dist}(\Psi(x), \Phi(x)), \quad y \in N(x, r)$$

there exist such $\xi = \xi(x, y, r)$ and $\eta = \eta(x, y, r)$ that

$$\Psi(\xi) \bigcap \Phi(\xi) \neq \emptyset, \quad \rho_X(\xi, x) \leq \frac{r}{\alpha - \beta} + \varepsilon \tag{24.14}$$

and, moreover,

$$\eta \in \Psi(\xi) \bigcap \Phi(\xi), \quad \rho_Y(\eta, y) \leq \beta \frac{r}{\alpha - \beta} + \varepsilon. \tag{24.15}$$

Proof. Again, dividing the metric ρ_Y over α, we will assume, without loss of generality, that $\alpha = 1$ and hence $\beta < 1$. Fix arbitrary $x \in X$, $r > \operatorname{dist}(\Psi(x), \Phi(x))$ and $y \in N(x, r)$.

4 The graphs gph (Ψ) and gph (Φ) are considered the subspaces of the metric space $X \times Y$ with the above defined metric on it (see (17.4)).

Choose $\delta > 0$ such that

$$\begin{aligned} &\beta + \delta < 1, \quad \frac{r}{1-(\beta+\delta)} \le \frac{r}{1-\beta} + \varepsilon, \\ &(\beta+\delta)\frac{r}{1-(\beta+\delta)} \le \beta\frac{r}{1-\beta} + \varepsilon. \end{aligned} \tag{24.16}$$

Take sequences $\{x_i\}$ and $\{y_i\}$ satisfying the statement of Lemma 24.13 for $x_0 = x$, $y_1 = y$. By virtue of (24.12) and (24.13) they are the Cauchy sequences. First let the subspace gph (Ψ) be complete. Then the mentioned sequences converge to points ξ and η respectively and, moreover, $\eta \in \Psi(\xi)$, since, by virtue of (24.11) we have $(x_i, y_i) \in$ gph (Ψ) $\forall i$. Passing in (24.11) to the limit as $i \to \infty$ and using the continuity of the set-valued map Φ, we get $\eta \in \Phi(\xi)$.

Now, let the space gph (Φ) be complete. Repeating previous reasoning for the set gph (Φ), we have $\eta \in \Phi(\xi)$. Passing in (24.11) to the limit, by virtue of the closedness of the set gph (Ψ) we obtain $\eta \in \Psi(\xi)$. Thus, we proved that $\eta \in \Psi(\xi) \bigcap \Phi(\xi)$.

Further, repeating the reasoning made during the proof of formula (24.5) in Theorem 24.6 and taking into account inequality (24.16) and estimates (24.10), (24.12) for all i we get

$$\rho_X(x_i, x_0) \le \frac{r}{1-(\beta+\delta)}$$

and hence

$$\rho_X(x_i, x_0) \le \frac{r}{1-\beta} + \varepsilon.$$

Passing in the obtained inequality to the limit as $i \to \infty$, we obtain estimate (24.14). Similarly, we have

$$\rho_Y(y_i, y_1) \le \beta\frac{r}{1-\beta} + \varepsilon,$$

that yields the estimate from (24.15). □

From the proved theorem as a simple particular case for $X = Y$ and $\Psi \equiv x$ is the identity map, the fixed point principle for contractive set-valued maps follows (see [66]).

Let the space X be complete. Then a sufficient condition of the completeness of the graph gph (Ψ) is its closedness and the completeness of the space Y, and a sufficient condition of the completeness of gph (Φ) is the "single-valuedness" of the set-valued map Φ, i.e., in this case for each x the set $\Phi(x)$ consists of a single point. Therefore, taking into account Remark 24.14, we conclude that Theorem 24.6 is the corollary of Theorem 24.15 for $r = \operatorname{dist}(\Psi(x), \Phi(x))$.

Moreover, by virtue of Remark 24.14, if for all $x_1, x_2 \in X$ and $y \in Y$ the infimums in the expressions $\operatorname{dist}(y, \Psi(x_1))$, $\operatorname{dist}(\Psi(x_1), \Phi(x_1))$ and $h(\Phi(x_1), \Phi(x_2))$ are achieved, then in the statement of Theorem 24.15 we can take $\varepsilon = 0$, $r = \operatorname{dist}(\Psi(x), \Phi(x))$, i.e., the following estimates hold true:

$$\begin{aligned} \rho_X(\xi, x) &\le \frac{\operatorname{dist}(\Psi(x), \Phi(x))}{\alpha - \beta}, \\ \rho_Y(\eta, y) &\le \beta\frac{\operatorname{dist}(\Psi(x), \Phi(x))}{\alpha - \beta}. \end{aligned} \tag{24.17}$$

Let us consider an example, demonstrating that in Theorem 24.15 it is impossible, generally speaking, to omit the assumption about the completeness of at least one of the graphs.

Example 24.16. Let X be the interval $[0, 1]$ with the natural metric, K the square on the plane $\mathbb{R}^2$ with the vertices $(0, 0)$, $(0, 1)$, $(1, 1)$, $(1, 0)$, and $\tilde{Y}$ a metric space consisting of the set of points $y = (y^1, y^2) \in K$ with the natural metric. For $x \in X$ denote by $\tilde{\Psi}(x) \subset K$ the vertical interval connecting the points $(x, 0)$ and $(x, 1)$, and by $\tilde{\Phi}(x) \subset K$ the interval connecting the points $(0, 0)$ and $(1, x/2)$. For each x the intervals $\tilde{\Psi}(x)$ and $\tilde{\Phi}(x)$ intersect in the unique point $y = (x, x^2/2)$ and hence the set of all points of intersection Π represents a part of parabola $y^2 = (y^1)^2/2$, $0 \le y^1 \le 1$. Set

$$Y = \tilde{Y} \setminus \Pi, \quad \Psi(x) = \tilde{\Psi}(x) \setminus \{(x, x^2/2)\}, \quad \Phi(x) = \tilde{\Phi}(x) \setminus \{(x, x^2/2)\}.$$

The so defined set-valued map Ψ is 1-covering and closed and the set-valued map Φ is Lipschitz with the constant $\beta = 1/2$. At the same time, obviously, the intersection $\Psi(x) \bigcap \Phi(x)$ is empty for all $x \in X$ (since, by construction we "deleted" from $\tilde{Y}$ all the points of intersection). The case is that both graphs gph (Ψ) and gph (Φ) (as well as the space Y itself) are not complete.

Nevertheless, notice that omitting the assumption of completeness of at least one of the graphs we can guarantee a more weak relation than (24.14).

Theorem 24.17. *Let $\alpha > \beta$ be given. Suppose that the space X is complete, a set-valued map Ψ is α-covering and sequentially upper semicontinuous and a set-valued map Φ is β-Lipschitz.*

Then for arbitrary $x \in X$ and $\varepsilon > 0$ there exists such a $\xi = \xi(x)$, that

$$\operatorname{dist}(\Psi(\xi), \Phi(\xi)) = 0, \quad \rho_X(\xi, x) \le \frac{\operatorname{dist}(\Psi(x), \Phi(x))}{\alpha - \beta} + \varepsilon.$$

The proof of this theorem is also based on Lemma 24.13 and is analogous to the proof of Theorem 24.15.

Let us compare Theorems 24.15 and 24.17. The main difference is that if at least one of the closed sets $\Psi(\xi)$ or $\Phi(\xi)$ is noncompact, then the relation $\operatorname{dist}(\Psi(\xi), \Phi(\xi)) = 0$ does not guarantee that their intersection is nonempty, i.e., that ξ is the coincidence point of the set-valued maps Ψ and Φ. To see this, we can consider the maps from Example 24.16. They satisfy all assumptions of Theorem 24.17 (the map Ψ is obviously sequentially upper semicontinuous) and $\operatorname{dist}(\Psi(x), \Phi(x)) = 0$ for all $x \in X$. At the same time, as mentioned, the set-valued maps Ψ and Φ have no coincidence points.

If we compare the assumptions of these theorems we see that in Theorem 24.17 for the map Ψ instead of the closedness we assume a stronger supposition, namely its sequential upper semicontinuity, but we do not suppose the completeness of any of the graphs gph (Ψ) and gph (Φ). But at the same time the statement of Theorem 24.17 is weaker than that of Theorem 24.15.

In this connection, the following question arises: is it possible to weaken the conditions of Theorem 24.17 by replacing the assumption of the sequential upper semicontinuity of the map Ψ with its closedness? The following example gives the negative answer to this question.

Example 24.18. Let K be the square defined in Example 24.16. Consider the metric spaces $X = \tilde{Y} = K$ with the natural metric. Their points will be denoted by $x = (x^1, x^2)$ and $y = (y^1, y^2)$ respectively. For $x \in X$ by $\tilde{\Phi}(x) \subset \tilde{Y}$ denote the interval $\{ y \colon y^2 = \frac{x^1}{2} y^1,\ 0 \le y^1 \le 1 \}$.

Set $\Pi_X = \{ x \colon x^2 = (x^1)^2/2,\ 0 \le x^1 \le 1 \}$. It is evident that $\Pi_X = \{ x \in X \colon x \in \tilde{\Phi}(x) \}$. Take in $\mathbb{R}^2$ a point $\bar{y}$ whose distance from K is greater than $\sqrt{2}$ (it is the diameter of the set K), e.g., $\bar{y} = (3, 3)$. Set

$$Y = \{ \bar{y} \} \cup (\tilde{Y} \setminus \Pi), \quad \Phi(x) \equiv \tilde{\Phi}(x) \setminus \{ (x^1, (x^1)^2/2) \}\ x \in X,$$

$$\Psi(x) = \begin{cases} x, & x \in X \setminus \Pi_X, \\ \bar{y}, & x \in \Pi_X. \end{cases}$$

(Here the set Π is defined as in Example 24.16.) In Y we consider the natural metric induced from $\mathbb{R}^2$. The so defined set-valued maps have closed values, a map Ψ is single-valued, 1-covering[5] and its graph is closed, and the set-valued map Φ is Lipschitz with the constant $\beta = 1/2$. At the same time $\operatorname{dist}(\Psi(x), \Phi(x)) > 0\ \forall x \in X$, since, by the single-valuedness of the map Ψ, the condition $\operatorname{dist}(\Psi(x), \Phi(x)) = 0$ is equivalent to $\Psi(x) \in \Phi(x)$, but the latter is impossible. Indeed, if $x \in \Pi_X$, then $\Psi(x) = \bar{y} \notin K$, but $\Phi(x) \subset K$, and if $x \notin \Pi_X$, then $\Psi(x) = x$, $x \notin \Phi(x)$ and hence $\Psi(x) \notin \Phi(x)$. The case is that although the map Ψ is closed, it is however not sequentially upper semicontinuous at the points $x \in \Pi_X$.

A local version of this coincidence theory is presented in [67].

5 To prove the 1-covering property we can use that if $x \in \Pi_X$, then $B^Y(\Psi(x), r) = \{ \bar{y} \}\ \forall r \le \sqrt{2}$, $\Psi(B^X(x, \sqrt{2})) = Y$.

25 Stability of coincidence points and properties of covering maps

Let us return to a question that is very important for applications, on stability of coincidence points of maps with respect to small perturbations which we started to discuss earlier.

We will use notation and notions introduced in the previous section.

Theorem 25.1. *Let $x_0 \in X$ be a coincidence point of set-valued maps Ψ and Φ. Let numbers $\alpha > \beta \geq 0$ be given. Suppose that sequences of set-valued maps $\{\Psi_n\}$, $\{\Phi_n\}$ are such that for each n the map Ψ_n is closed and α-covering and the map Φ_n is β-Lipschitz and at least one of the graphs* $\operatorname{gph}(\Psi_n)$ *or* $\operatorname{gph}(\Phi_n)$ *is a complete set. Assume also that*

$$h^+(\Psi(x_0), \Psi_n(x_0)) \to 0, \quad h^+(\Phi(x_0), \Phi_n(x_0)) \to 0, \quad n \to \infty. \tag{25.1}$$

Then for each converging to zero sequence of positive numbers $\{\delta_n\}$ there exists such a sequence $\{x_n\}$ that

$$\Psi_n(x_n) \bigcap \Phi_n(x_n) \neq \emptyset \quad \forall n, \tag{25.2}$$

$$\rho_X(x_n, x_0) \to 0, \quad n \to \infty, \tag{25.3}$$

and, moreover,

$$\rho_X(x_n, x_0) \leq \frac{h^+(\Psi(x_0), \Psi_n(x_0)) + h^+(\Phi(x_0), \Phi_n(x_0))}{\alpha - \beta} + \delta_n. \tag{25.4}$$

Proof. Fix a number n and apply Theorem 24.15 to maps Ψ_n and Φ_n at the point $x = x_0$. By virtue of this theorem there exist such $x_n \in X$ that we have (25.2) and

$$\rho_X(x_n, x_0) \leq \frac{\operatorname{dist}(\Psi_n(x_0), \Phi_n(x_0))}{(\alpha - \beta)} + \delta_n. \tag{25.5}$$

Applying inequality (15.8) twice, we get

$$\begin{aligned}\operatorname{dist}(\Psi_n(x_0), \Phi_n(x_0)) &\leq \operatorname{dist}(\Psi_n(x_0), \Phi(x_0)) + h^+(\Phi(x_0), \Phi_n(x_0)) \\ &\leq \operatorname{dist}(\Psi(x_0), \Phi(x_0)) + h^+(\Psi(x_0), \Psi_n(x_0)) + h^+(\Phi(x_0), \Phi_n(x_0)) \\ &= h^+(\Psi(x_0), \Psi_n(x_0)) + h^+(\Phi(x_0), \Phi_n(x_0)).\end{aligned}$$

Here the last equality follows from the fact that, by assumption, $\Psi(x_0) \bigcap \Phi(x_0) \neq \emptyset$ and hence $\operatorname{dist}(\Psi(x_0), \Phi(x_0)) = 0$. Thus we have

$$\operatorname{dist}(\Psi_n(x_0), \Phi_n(x_0)) \leq h^+(\Psi(x_0), \Psi_n(x_0)) + h^+(\Phi(x_0), \Phi_n(x_0)).$$

From the obtained inequality, by virtue of (25.5) we obtain (25.4), and hence (25.3) also. ☐

DOI 10.1515/9783110460308-026

Notice that Theorem 25.1 is the strengthening of Theorem 2 from [65].

Let us clear up the question of when the limit of a convergent sequence of α-covering set-valued maps is covering. As we remember from Example 24.4, the pointwise limit of covering set-valued maps may be noncovering. However, if we strengthen the assumption about convergence, we can get the following assertion.

Theorem 25.2. *Suppose that a space X is complete and for each bounded set $U \subseteq X$ the uniform convergence holds true:*

$$h(\Psi_n(x), \Psi(x)) \rightrightarrows 0, \quad n \to \infty, \; x \in U. \tag{25.6}$$

Then for an arbitrary $\varepsilon \in (0, \alpha)$ the map Ψ is $(\alpha - \varepsilon)$-covering.

In brief, this theorem states that the uniform limit of α-covering maps is $(\alpha - \varepsilon)$-covering for each $0 < \varepsilon < \alpha$.

The proof of the theorem is based on the following assertion.

Lemma 25.3. *Let a space X be complete and a set-valued map Ψ be closed. Suppose that there exists $\alpha > 0$ for which the following holds:*

$$\mathrm{cl}\,(\Psi(O^X(x, r))) \supseteq O^Y(\Psi(x), \alpha r) \quad \forall r > 0, \; \forall x \in X. \tag{25.7}$$

Then for each $\delta > 0$ we get

$$\Psi(O^X(x, r + \delta)) \supseteq \mathrm{cl}\,(\Psi(O^X(x, r))) \quad \forall r > 0, \; \forall x \in X. \tag{25.8}$$

This statement is a particular case of Lemma 36 from [55], Ch. 6 and also follows from Lemma 25.5 that will be proved below.

Proof of Theorem 25.2. Take an arbitrary positive $\gamma < \alpha$ and prove the inclusion

$$\mathrm{cl}\,(\Psi(O^X(x, r))) \supseteq O^Y(\Psi(x), \gamma r) \quad \forall r > 0, \; \forall x \in X. \tag{25.9}$$

Indeed, fix arbitrary $x \in X$, $r > 0$ and let $y \in O^Y(\Psi(x), \gamma r)$. Let us show that

$$y \in \mathrm{cl}\,(\Psi(O^X(x, r))). \tag{25.10}$$

Take an arbitrary $\eta > 0$, for which $\eta + \gamma r < \alpha r$. By virtue of the choice of the point y, there exists $y_1 \in \Psi(x)$ such that $\rho_Y(y_1, y) \le \gamma r$. From the assumptions of the theorem it follows that there exists a number n, for which the following inclusions hold:

$$O^Y(\Psi_n(x), \eta) \supseteq \Psi(x), \quad O^Y(\Psi(\xi), \eta) \supseteq \Psi_n(\xi) \quad \forall \xi : \rho_X(\xi, x) \le r.$$

By the first of these inclusions there exists $y_2 \in \Psi_n(x)$, for which $\rho_Y(y_2, y_1) \le \eta$. Taking into account the triangle inequality and the choice of the number η we get $\rho_Y(y_2, y) \le \eta + \gamma r < \alpha r$. Since the map Ψ_n is α-covering, there exists $x_n \in X$ such that $y \in \Psi_n(x_n)$ and $\rho_X(x, x_n) < r$ and hence $x_n \in O^X(x, r)$. Thus, by virtue of the second inclusion for $\xi = x_n$ we have $y \in O^Y(\Psi(x_n), \eta))$ and therefore $\operatorname{dist}(y, \Psi(O^X(x, r))) \le \eta$. By the arbitrariness of η we obtain (25.10).

So, relation (25.9) is proved. Then, by Lemma 25.3 (see (25.8) and (25.9)) we have

$$\Psi(O^X(x, r + \delta)) \supseteq O^Y(\Psi(x), \gamma r) \quad \forall r > 0,\ \forall x \in X,\ \forall \delta > 0.$$

Take an arbitrary $\varepsilon \in (0, \alpha)$. From the obtained inclusion which is true for every $\gamma < \alpha$, taking $\gamma = \alpha - \varepsilon/4$ we have

$$\Psi(O^X(x, r)) \supseteq O^Y(\Psi(x), (\alpha - \varepsilon/2)r) \quad \forall r > 0,\ \forall x \in X.$$

From here, by the obvious inclusion

$$O^Y(\Psi(x), (\alpha - \varepsilon/2)r) \supseteq B^Y(\Psi(x), (\alpha - \varepsilon)r)$$

we finally obtain

$$\Psi(O^X(x, r)) \supseteq B^Y(\Psi(x), (\alpha - \varepsilon)r) \quad \forall r > 0,\ \forall x \in X.$$

Thus, the map Ψ is $(\alpha - \varepsilon)$-covering, which concludes the proof of the theorem. □

The following example demonstrates that, generally speaking, we cannot take $\varepsilon = 0$ in the proved theorem.

Example 25.4. We will use the construction of Example 24.8. Similarly to it, take $X = l_1$, $Y = \mathbb{R}$ and for $x \in l_1$ set $\Psi(x) = \langle x^*, x \rangle$, where $x^* \in l_\infty$ is defined in (24.7). For natural n and $x \in l_1$ set $\Psi_n(x) = \Psi(x) + n^{-1}\langle e, x \rangle$, where $e = (1, \dots, 1, \dots) \in l_\infty$. Then the sequence Ψ_n uniformly converges to Ψ on each bounded set, all maps Ψ_n are 1-covering, however the map Ψ is not 1-covering.

The following important assertion has various applications.

Lemma 25.5. *Suppose that a set-valued map Ψ has a complete graph and there exists $\alpha > 0$, for which* (25.7) *holds. Then for an arbitrary $\delta > 0$ relation* (25.8) *holds true.*

Proof. Fix $x_0 \in X$, $r_0, \delta > 0$ and take an arbitrary $y \in \operatorname{cl}(\Psi(O^X(x_0, r_0)))$. We should prove

$$y \in \Psi(O^X(x_0, r_0 + \delta)). \tag{25.11}$$

Let us do it.

Take a convergent series with positive members γ_i such that the sequence $\{\gamma_i\}$ monotonically decreases and

$$2\alpha^{-1} \sum_{i=1}^{\infty} \gamma_i < \delta.$$

Take in Y a sequence $\{\tilde{y}_i\}$ such that $\tilde{y}_1 = y$, $\rho_Y(\tilde{y}_i, y) < \gamma_i/2\ \forall i$. From the triangle inequality and the relation $\gamma_{i+1} < \gamma_i$ it follows that $\rho_Y(\tilde{y}_i, \tilde{y}_{i+1}) < \gamma_i\ \forall i$.

Successively construct the pairs of points $(y_1, x_1), (y_2, x_2), \ldots$, where $x_i \in X$, $y_i \in Y$, in the following way. By construction, $\tilde{y}_1 \in \mathrm{cl}\,(\Psi(O^X(x_0, r_0)))$. Hence there exists $y_1 \in \Psi(O^X(x_0, r_0))$ such that $\rho_Y(y_1, \tilde{y}_1) < \gamma_1$. Therefore, there exists $x_1 \in X$ such that

$$\Psi(x_1) = y_1, \quad \rho_X(x_0, x_1) < r_0.$$

Let us turn to the construction of the pair (y_2, x_2). We have

$$\rho_Y(\Psi(x_1), \tilde{y}_2) = \rho_Y(y_1, \tilde{y}_2) \le \rho_Y(y_1, \tilde{y}_1) + \rho_Y(\tilde{y}_1, \tilde{y}_2) < \gamma_1 + \gamma_1 = 2\gamma_1.$$

Thus $\tilde{y}_2 \in O^Y(\Psi(x_1), 2\gamma_1)$, from which, by (25.7):

$$\tilde{y}_2 \in \mathrm{cl}\,(\Psi(O^X(x_1, 2\alpha^{-1}\gamma_1))).$$

Therefore there exists $y_2 \in Y$ such that

$$y_2 \in \Psi(O^X(x_1, 2\alpha^{-1}\gamma_1)), \quad \rho_Y(y_2, \tilde{y}_2) < \gamma_2.$$

Hence there exists $x_2 \in X$ such that

$$\Psi(x_2) = y_2, \quad \rho_X(x_1, x_2) < 2\alpha^{-1}\gamma_1.$$

Now, let us turn to the construction of the pair (y_3, x_3). We have

$$\rho_Y(\Psi(x_2), \tilde{y}_3) = \rho_Y(y_2, \tilde{y}_3) \le \rho_Y(y_2, \tilde{y}_2) + \rho_Y(\tilde{y}_2, \tilde{y}_3) < 2\gamma_2,$$

from which, by (25.7)

$$\tilde{y}_3 \in \mathrm{cl}\,(\Psi(O^X(x_2, 2\alpha^{-1}\gamma_2))).$$

Hence there exists $y_3 \in Y$ such that

$$y_3 \in \Psi(O^X(x_2, 2\alpha^{-1}\gamma_2)), \quad \rho_Y(y_3, \tilde{y}_3) < \gamma_3.$$

Therefore, there exists $x_3 \in X$ such that

$$\Psi(x_3) = y_3, \quad \rho_X(x_2, x_3) < 2\alpha^{-1}\gamma_2.$$

Continuing this process, we successfully construct the pairs (y_i, x_i), $i = 1, 2, \ldots$, for which we have

$$\Psi(x_i) = y_i, \quad \rho_X(x_i, x_{i+1}) < 2\alpha^{-1}\gamma_i, \quad \rho_Y(y_i, \tilde{y}_i) < \gamma_i.$$

In the Cartesian product $X \times Y$ consider the sequence $\{(x_i, y_i)\}$. Evidently, it is the Cauchy sequence and lies in the graph $\mathrm{gph}\,\Psi$. Since this graph, by assumption, is a complete space, the sequence converges to a certain point $(\bar{x}, \bar{y}) \in \mathrm{gph}\,\Psi$. So, $\bar{y} \in \Psi(\bar{x})$. In addition, $x_i \to \bar{x}$, $y_i \to \bar{y}$, and by construction $y_i \to y$ and hence $y = \bar{y}$ and $y \in \Psi(\bar{x})$. In addition, from the triangle inequality and the choice of the sequence $\{y_i\}$ it easily follows that

$$\rho_X(x_0, \bar{x}) \le \sum_{i=0}^{\infty} \rho_X(x_i, x_{i+1}) < r_0 + 2\alpha^{-1} \sum_{i=1}^{\infty} \gamma_i < r_0 + \delta.$$

Therefore, $\rho_X(x_0, \bar{x}) < r_0 + \delta$ and hence $y \in \Psi(O^X(x_0, r_0 + \delta))$, completing the proof of (25.11). □

The validity of Lemma 25.3 follows from Lemma 25.5, since if the space X is complete and the map Ψ is closed, then its graph is complete. From Lemma 25.5 also immediately follows the next assertion.

Corollary 25.6. *Let the graph of a set-valued map Ψ be complete. Suppose that there exists $\alpha > 0$ for which (25.7) holds. Then for an arbitrary $\varepsilon \in (0, \alpha)$, the map Ψ is $(\alpha - \varepsilon)$-covering.*

Notice that in Lemma 25.5 and its corollary the assumption on the completeness of the graph of Ψ is essential and it cannot be omitted.

The class of maps satisfying (25.7) for a certain $\alpha > 0$ was introduced and studied by B. A. Pasynkov who called them *α-filling maps*. In terms of α-filling maps the above corollary may be reformulated in the following form.

Theorem 25.7. *For a given $\alpha > 0$, each α-filling set-valued map with the complete graph is $(\alpha - \varepsilon)$-covering for an arbitrary $\varepsilon \in (0, \alpha)$.*

In conclusion, let us discuss the structure of the set of coincidence points Ξ of maps Ψ and Φ. Namely, denote

$$\Xi = \{\xi \in X : \Psi(\xi) \cap \Phi(\xi) \neq \emptyset\}$$

and for $x \in X$ by $\Xi(x)$ denote the set (possibly, empty) of such points $\xi \in \Xi$, for which (24.17) holds.

Notice at once that if for each $\varepsilon > 0$ there exists $\xi \in X$, for which (24.14) holds then we have

$$\Xi \neq \emptyset, \quad \operatorname{dist}(x, \Xi) \leq \frac{\operatorname{dist}(\Psi(x), \Phi(x))}{\alpha - \beta}.$$

The sets Ξ and $\Xi(x)$ may be arranged very "poorly". In particular, the following example shows that the set Ξ and the set $\Xi(x)$ for a certain x may turn out to be nonclosed.

Example 25.8. Let $X = [0, 1]$, $Y = \{y \in \mathbb{R} : y \geq 1\} \cup \{0\}$, $\Psi(x) = \{x^{-1}\}$ $\forall x \in (0, 1]$, $\Psi(0) = \{0\}$, $\Phi(x) \equiv \{y \in \mathbb{R} : y \geq 1\}$. It is obvious that the set-valued map Ψ is closed and 1-covering. It is easy to see that

$$\Xi = (0, 1], \quad \Xi(x) = \{x\}\ \forall x \in (0, 1], \quad \Xi(0) = (0, 1]$$

and hence the sets Ξ and $\Xi(0)$ are nonclosed.

At the same time, the next lemma shows that under additional assumptions concerning the space X and the maps Ψ, Φ, we can guarantee that the set-valued map $\Xi(\cdot)$ is sequentially upper semicontinuous. Let $\alpha > \beta \geq 0$.

Lemma 25.9. *Suppose that the set-valued maps Ψ and Φ are continuous and, moreover, the map Φ is compact-valued. Let each closed bounded set in the space X be compact.*

Then the set $\tilde{X} = \{x \in X : \Xi(x) \neq \emptyset\}$ is closed and the map $\Xi(\cdot)$ is sequentially upper semicontinuous on it.

Proof. Let $x_i \in \tilde{X}$, $x_i \to x_0$ and $\xi_i \in \Xi(x_i)$ for all i. Let us show that then the set $\Xi(x_0)$ is nonempty and $\operatorname{dist}(\xi_i, \Xi(x_0)) \to 0$. Suppose the contrary. Then there exists such a $\delta > 0$, that, passing to a subsequence if necessary, we get $\operatorname{dist}(\xi_i, \Xi(x_0)) \geq \delta$ for all i (we assume that the distance to the empty set equals infinity).

Since the set-valued maps Ψ, Φ are continuous, the number sequence $\{\operatorname{dist}(\Psi(x_i), \Phi(x_i))\}$ is bounded. By the inequalities

$$\rho_X(x_i, \xi_i) \leq \frac{\operatorname{dist}(\Psi(x_i), \Phi(x_i))}{\alpha - \beta} \tag{25.12}$$

the sequence $\{\rho_X(x_i, \xi_i)\}$ is also bounded. Thus the sequence $\{\xi_i\}$ is bounded and passing to a subsequence, if necessary, we get $\xi_i \to \xi_0$.

For each i choose $y_i \in \Psi(\xi_i) \bigcap \Phi(\xi_i)$. Then, by the sequential upper semicontinuity of the map Φ we have $\operatorname{dist}(y_i, \Phi(\xi_0)) \to 0$, from which, by using the compactness of the set $\Phi(\xi_0)$, we get, without loss of generality that $y_i \to y_0 \in \Phi(\xi_0)$. Hence, by virtue of the closedness of the map Ψ we have $y_0 \in \Psi(\xi_0)$ and so $y_0 \in \Psi(\xi_0) \bigcap \Phi(\xi_0) \neq \emptyset$, implying $\xi_0 \in \Xi$. We see that $\xi_0 \in \Xi(x_0)$, since passing to the limit in (25.12) as $i \to \infty$, by the continuity of Ψ and Φ we have

$$\rho_X(x_0, \xi_0) \leq \operatorname{dist}(\Psi(x_0), \Phi(x_0))/(\alpha - \beta).$$

So, we proved that $\operatorname{dist}(\xi_i, \Xi(x_0)) \to 0$, contradicting to the existence of $\delta > 0$ chosen earlier. □

Notice that the maps from Example 24.7 satisfy all conditions of the lemma, but the corresponding map $\Xi(\cdot)$ is not lower semicontinuous.

26 Topological degree and fixed points of set-valued maps in Banach spaces

Topological degree theory for set-valued maps with convex values is a powerful and convenient tool for obtaining fixed point results for maps in normed and more general classes of topological vector spaces (e.g., in locally convex spaces). These principles have another nature than the Banach fixed point theorem on contractive maps. Restricting ourselves for simplicity to the maps in Banach spaces, in this section we will give an outline of this theory. Details can be found in [17, 21, 22].

26.1 Topological degree of single-valued maps

We will start with the draft of the classical Brouwer[1] degree theory in finite-dimensional spaces. In our exposition we follow mainly the book of J. Milnor [68].

Let $U \subset \mathbb{R}^k$ and $V \subset \mathbb{R}^l$ be open sets, and $\varphi\colon U \to V$ a continuous map. The map φ is said to be *smooth* if all partial derivatives of its components

$$\frac{\partial^\alpha \varphi_j(x_1, \dots, x_m)}{\partial x_{i_1} \cdots \partial x_{i_\alpha}}$$

exist and are continuous.

If $X \subset \mathbb{R}^k$ and $Y \subset \mathbb{R}^l$ are arbitrary subsets, then a map $\varphi\colon X \to Y$ is called *smooth* if for each point $x \in X$ there exists an open neighborhood $U(x) \subset \mathbb{R}^k$ and a smooth map $\tilde{\varphi}\colon U(x) \to \mathbb{R}^l$ coinciding with φ on $U(x) \cap X$.

A map $\varphi\colon X \to Y$ is called *the homeomorphism* of sets X and Y if $\varphi(X) = Y$, the map φ is one-to-one (i.e., $x \neq y$ implies $\varphi(x) \neq \varphi(y)$) and both the maps φ and φ^{-1} are continuous. A homeomorphism $\varphi\colon X \to Y$ is called *diffeomorphism* provided φ and φ^{-1} are smooth maps.

A subset $M \subset \mathbb{R}^k$ is called *a smooth manifold of dimension m (or smooth m-dimensional manifold)* if for each point $x \in M$ there exists a neighborhood $W \cap M$ which is diffeomorphic to an open subset $U \subset \mathbb{R}^m$.

A diffeomorphism $g\colon U \to W \cap M$ is called *a parametrization of the domain $W \cap M$* and the inverse diffeomorphism $g^{-1}\colon W \cap M \to U$ is called *a system of coordinates on $W \cap M$*.

Now, let us consider the notions of the derivative of a map and the tangent space to a manifold.

1 Luitzen Egbertus Jan Brouwer (1881–1966), a Dutch mathematician.

DOI 10.1515/9783110460308-027

If $U \subset \mathbb{R}^k$ is an open set and $\varphi: U \to \mathbb{R}^l$ is a smooth map then its derivative at a point $x \in U$ is a linear map $\varphi'_x: \mathbb{R}^k \to \mathbb{R}^l$ given by the matrix

$$\varphi'_x = \begin{pmatrix} \frac{\partial\varphi_1(x_1,\ldots,x_k)}{\partial x_1} & \cdots & \frac{\partial\varphi_1(x_1,\ldots,x_k)}{\partial x_k} \\ \vdots & \ddots & \vdots \\ \frac{\partial\varphi_l(x_1,\ldots,x_k)}{\partial x_1} & \cdots & \frac{\partial\varphi_l(x_1,\ldots,x_k)}{\partial x_1} \end{pmatrix}$$

Now we can define *the tangent space* TM_x for a smooth m-dimensional manifold $M \subset \mathbb{R}^k$ at a point $x \in M$. Take a parametrization $g: U \to M \subset \mathbb{R}^k$ of a neighborhood $g(U)$ of a point x, $g(u) = x$, where U is an open subset of $\mathbb{R}^m$. Let $g'_u: \mathbb{R}^m \to \mathbb{R}^k$ be the derivative of g at the point $u \in U$. Define TM_x as the image $g'_u(\mathbb{R}^m)$.

Geometrically, we can imagine the m-dimensional plane in $\mathbb{R}^k$ which approximates M in the best way in a neighborhood of x. Then TM_x is the parallel plane passing through the origin.

Now, let $M \subset \mathbb{R}^k$ and $N \subset \mathbb{R}^l$ be smooth manifolds of dimensions m and n respectively and $\varphi: M \to N$ a smooth map. Take points $x \in M$ and $y = \varphi(x) \in N$. Our purpose is to define the derivative φ'_x as the linear map

$$\varphi'_x: TM_x \to TN_y.$$

Since φ is a smooth map, there exists an open neighborhood W of the point x and a smooth map $\tilde{\varphi}: W \to \mathbb{R}^l$ coinciding with φ on $W \cap M$. Define $\varphi'_x(v) = \tilde{\varphi}'_x(v)$ for $v \in TM_x$. Let us show that the image $\tilde{\varphi}'_x(v)$ belongs to the space TN_y and it does not depend on the choice of the extension $\tilde{\varphi}$.

Choose parametrizations $g: U \to M \subset \mathbb{R}^k$ and $h: V \to N \subset \mathbb{R}^l$ of neighborhoods $g(U)$ and $h(V)$ of points x and y respectively. Diminishing U, if necessary, we can assume that $g(U) \subset W$ and $\tilde{\varphi}(g(U)) \subset h(V)$. Thus the smooth map $\hat{\varphi} = h^{-1} \circ \varphi \circ g: U \to V$ is well defined. Moreover, from the definition of $\tilde{\varphi}$ it follows that $\hat{\varphi} = h^{-1} \circ \tilde{\varphi} \circ g$.

Passing to the derivatives, we can represent $\hat{\varphi}'_u: \mathbb{R}^m \to \mathbb{R}^n$ in the form

$$\hat{\varphi}'_u = (h'_w)^{-1} \circ \tilde{\varphi}'_x \circ g'_u,$$

where $u = g^{-1}(x)$, $w = h^{-1}(y)$, $g'_u: \mathbb{R}^m \to \mathbb{R}^k$, $\tilde{\varphi}'_x: \mathbb{R}^k \to \mathbb{R}^l$, and $h'_w: \mathbb{R}^n \to \mathbb{R}^l$.

From here it follows that the map $\tilde{\varphi}'_x$ transforms $TM_x = Im(g'_u)$ into $TN_y = Im(h'_w)$. Moreover, since the map $\tilde{\varphi}'_x$ on TM_x may be represented as $h'_w \circ \hat{\varphi}'_u \circ (g'_u)^{-1}$ the derivative φ'_x does not depend on the choice of $\tilde{\varphi}$ and so it is well defined.

Let, as earlier, $\varphi: M \to N$ be a smooth map of manifolds M and N. A point $x \in M$ is said to be *regular* provided the derivative $\varphi'_x: TM_x \to TM_y$ has the maximal rank (i.e., $\min\{m, n\}$), otherwise the point x is called *singular*. The set of all singular points of φ is denoted by $\mathcal{S}_\varphi$. The set $\varphi(\mathcal{S}_\varphi)$ consists of singular values of φ. A point $y \in N$ is *the regular value of* φ if its pre-image $\varphi^{-1}(y)$ contains only regular points (in particular,

$\varphi^{-1}(y) = \emptyset$). The following assertion, which is of a great importance in differential topology and nonlinear analysis is called the Sard[2] theorem.

Theorem 26.1. *The set of singular values $\varphi(\mathcal{S}_\varphi)$ has a zero Lebesgue measure.*

The proof can be found, for example in [68].

Now recall the notion of orientation. Two bases $(e_1, \ldots, e_n)$ and $(e'_1, \ldots, e'_n)$ of the space $\mathbb{R}^n$ have the same orientation if the determinant of the matrix of transfer from one basis to another is positive. The orientation in $\mathbb{R}^n$ is defined by the choice of one of the two equivalence classes of bases with respect to this relation. The standard orientation in $\mathbb{R}^n$ is generated by the basis

$$\{1, 0, \ldots, 0\}, \ldots, \{0, \ldots, 0, 1\}.$$

A smooth manifold M is called *oriented* provided all its tangent spaces are concordantly oriented. The concordance of orientations means that each point of M possesses a neighborhood $\mathcal{U} \subset M$ and a diffeomorphism $h: \mathcal{U} \to \mathcal{V}$, where $\mathcal{V} \subset \mathbb{R}^m$ is an open set such that for each $x \in \mathcal{U}$ the isomorphism h'_x transforms the chosen orientation of the space TM_x into the standard orientation of the space $\mathbb{R}^m$.

Now, let M and N be two oriented smooth manifolds of the same dimension n, suppose that M is compact and N is arcwise connected, i.e., any two points in N may be connected by a continuous curve. Let $\varphi: M \to N$ be a smooth map and $y \in N$ its regular value (it exists by Theorem 26.1). Notice that the set $\varphi^{-1}(y)$ is finite. Indeed, if $x_0 \in \varphi^{-1}(y)$ then the Jacobian $\det \varphi'_{x_0}$ is nonzero and, by the classical Inverse Mapping Theorem, there exists a neighborhood $U(x_0)$ of the point x_0 such that φ maps it homeomorphically onto a certain neighborhood of y. It means that the point x_0 is isolated in the sense that $U(x_0) \cap \varphi^{-1}(y) = \{x_0\}$. So, if we suppose that the set $\varphi^{-1}(y)$ is infinite then, by virtue of compactness of M, it must have an accumulation point x_* which obviously also belongs to $\varphi^{-1}(y)$ but is not isolated, that gives the contradiction.

Let $x \in M$ be a regular point of a map φ and hence $\varphi'_x: TM_x \to TN_y$ is a linear isomorphism. Let *sign* φ'_x be equal to $+1$ or -1 depending on whether φ'_x preserves the orientation or not.

Define *the topological degree of φ at a regular value y* as

$$\deg(\varphi, y) = \sum_{x \in \varphi^{-1}(y)} \operatorname{sgn} \varphi'_x.$$

It can be shown (the proof may be found in [68]) that this integer does not depend on the choice of a regular value $y \in N$. It is called *the topological degree* $\deg \varphi$ *of a map* $\varphi: M \to N$. Notice that this characteristic is invariant with respect to smooth

2 Arthur Sard (1909–1980), an American mathematician.

deformations of a map φ in the following sense: if $\psi\colon M \times [0,1] \to N$ is a smooth map then $\deg \psi(\cdot, 0) = \deg \psi(\cdot, 1)$ (see [68]).

Let $\varphi\colon M \to N$ be a continuous (not necessarily smooth) map of the same manifolds. Consider a smooth ε-approximation φ_ε of the map φ, i.e., such a smooth map $\varphi_\varepsilon\colon M \to N$ for which

$$\|\varphi(x) - \varphi_\varepsilon(x)\| < \varepsilon \quad \text{for all } x \in M.$$

If $\varepsilon > 0$ is sufficiently small, the degree $\deg \varphi_\varepsilon$ does not depend on ε. It is called *the topological degree of a continuous map* φ. This characteristic is also invariant with respect to continuous deformations of φ which are called *homotopies* (see [68]).

Now, let $U \subset \mathbb{R}^n$ be an open bounded set and $\varphi\colon \operatorname{cl} U \to \mathbb{R}^n$ a continuous map. A zero point $x_0 \in U$, $\varphi(x_0) = 0$ of the map φ is called *isolated* if φ has no more zero points in a certain neighborhood of x_0. It means that $\varphi(x) \neq 0$ for all $x \in S(x_0, r)$, where $S(x_0, r)$ is a sphere centered at x_0 of a sufficiently small radius $r > 0$. Denote by $S^{n-1} \subset \mathbb{R}^n$ the unit sphere centered at the origin. From the invariance of degree with respect to continuous homotopies it easily follows (verify it as an exercise) that the topological degree $\deg \tilde{\varphi}$ of a map $\tilde{\varphi}\colon S(x_0, r) \to S^{n-1}$, $\tilde{\varphi}(x) = \varphi(x)/\|\varphi(x)\|$ does not depend on r, it is called *the topological index of* x_0 and is denoted as $\operatorname{ind} x_0$.

We need the following auxiliary assertion (see [69], Theorem 1.3).

Lemma 26.2. *Every continuous map $\varphi\colon \partial U \to \mathbb{R}^n$ such that $\varphi(x) \neq 0$ for all $x \in \partial U$ may be extended to a continuous map $\tilde{\varphi}\colon \operatorname{cl} U \to \mathbb{R}^n$ with a finite number of zero points. Moreover, if U is connected, φ may be extended with a unique zero point.*

Let $f\colon \partial U \to \mathbb{R}^n$ be a continuous map which is *fixed point free*, i.e., $x \neq f(x)$ for all $x \in \partial U$. It means that the vector field $\varphi\colon \partial U \to \mathbb{R}^n$ corresponding to f defined as $\varphi(x) = x - f(x)$ has no zero points (we will call such fields *nondegenerate*). Let $\tilde{\varphi}\colon \operatorname{cl} U \to \mathbb{R}^n$ be an extension of the field φ with a finite number of zero points $x_1, \dots, x_k$. The integer characteristic

$$\sum_{i=1}^{k} \operatorname{ind} x_i$$

does not depend on the choice of an extension $\tilde{\varphi}$ (see [70]). It is called *the topological degree*

$$\deg(\varphi, \partial U)$$

of the vector field $\varphi = i - f$ on the boundary ∂U.

Let us describe the main properties of the introduced characteristic (details can be found, for example, in [71, 18, 69, 72]).

(i) *Normalization property.* If $f(x) \equiv x_0$ for all $x \in \partial U$ then

$$\deg(\varphi, \partial U) = \begin{cases} 1, & \text{if } x_0 \in U, \\ 0, & \text{if } x_0 \notin U. \end{cases}$$

(ii) *Homotopy invariance.* Assume that vector fields $\varphi_0 = i - f_0$ and $\varphi_1 = i - f_1$ are homotopic ($\varphi_0 \sim \varphi_1$), i.e., there exists a continuous map $g\colon \partial U \times [0,1] \to \mathbb{R}^n$ such that
   (a) $x \neq g(x,\lambda)$ for all $x \in \partial U$ and $\lambda \in [0,1]$;
   (b) $g(\cdot,0) = f_0,\ g(\cdot,1) = f_1$.
   Then
$$\deg(\varphi_0, \partial U) = \deg(\varphi_1, \partial U).$$

The homotopy invariance principle is widely used for the calculation of the degree, allowing the reduction of the problem from complicated fields to more simple ones. In this connection concrete conditions of homotopy play a very important role. One of the most simple and geometrically clear conditions is expressed by the following *Poincare[3]–Bohl[4] theorem.*

Theorem 26.3. *Let nondegenerate vector fields $\varphi_0 = i - f_0$ and $\varphi_1 = i - f_1$ not admit opposite directions on ∂U, i.e.,*
$$\frac{\varphi_0(x)}{\|\varphi_0(x)\|} \neq \frac{\varphi_1(x)}{\|\varphi_1(x)\|} \quad \text{for all } x \in \partial U.$$
Then $\varphi_0 \sim \varphi_1$ and hence $\deg(\varphi_0, \partial U) = \deg(\varphi_1, \partial U)$.

Proof. It is easy to see that the desired homotopy is generated by the linear deformation $g\colon \partial U \times [0,1] \to \mathbb{R}^n$,
$$g(x,\lambda) = (1-\lambda) f_0(x) + \lambda f_1(x).$$ □

The following property of the degree is called *the additive dependence on the domain.*

(iii) Let $\{U_j\}_{j\in J}$ be a family of disjoint open subsets of U and a vector field φ is defined and nondegenerate on $\operatorname{cl} U \setminus \bigcup_{j\in J} U_j$. Then the degrees $\deg(\varphi, \partial U_j)$ are nonzero for only finitely many indices $j \in J$ and
$$\deg(\varphi, \partial U) = \sum_{j\in J} \deg(\varphi, \partial U_j).$$

It should be mentioned that the above properties (i)–(iii) of the topological degree are of a basic character in the sense that other assertions of topological degree theory may be deduced from them. As an example, let us consider the following general fixed point principle which is of great importance for us.

Theorem 26.4. *Let a continuous map $f\colon \operatorname{cl} U \to \mathbb{R}^n$ be fixed point free on the boundary ∂U and* $\deg(i - f, \partial U) \neq 0$. *Then f admits a fixed point $x_* \in U$, $x_* = f(x_*)$.*

3 Jules Henri Poincare (1854–1912), a French mathematician, physicist, and philosopher.
4 Piers Bohl (1865–1921), a Latvian mathematician.

Proof. Suppose, to the contrary, that f is fixed point free on the whole cl U. Then, since the set cl U is compact it is possible to represent cl U as the union of a finite number of closures cl U_j of disjoint open sets U_j such that for each $x, y \in \operatorname{cl} U_j$ the vectors $\varphi(x)$ and $\varphi(y)$, where $\varphi = i - f$, form an acute angle. Then, for each given set U_{j_0} we may find a vector $x_0 \notin \operatorname{cl} U_{j_0}$ such that for each $x \in \operatorname{cl} U_{j_0}$ the vectors $\varphi(x)$ and $\psi(x) = x - x_0$ form an acute angle. By Theorem 26.3 the fields $\varphi(x)$ and $\psi(x)$ are homotopic on ∂U_{j_0} and hence, applying properties (i) and (ii) we get

$$\deg(\varphi, \partial U_{j_0}) = \deg(\psi, \partial U_{j_0}) = 0.$$

Property (iii) yields $\deg(\varphi, \partial U) = 0$ contradicting to the assumption. □

This principle may be used to provide simple and geometrically clear proofs of a number of fixed point results. As an example, we apply it to obtain the classical *Brouwer fixed point theorem.*

Theorem 26.5. *Let $\mathcal{M}$ be a nonempty convex compact (i.e., closed and bounded) subset of $\mathbb{R}^n$, $n \geq 1$. Then each continuous map $f: \mathcal{M} \to \mathcal{M}$ has a fixed point.*

The only additional tool that we will exploit in the proof is the following retraction result which is the consequence of the known Tietze[5]–Dugundji[6] extension theorem (see, e.g., [72]).

Lemma 26.6. *Every closed convex subset $\mathcal{M}$ of a normed space X is a retract of X, i.e., there exists a continuous map (retraction) $\varrho: X \to \mathcal{M}$ such that $\varrho(x) = x\ \forall x \in \mathcal{M}$.*

Remark 26.7. If the space X is finite-dimensional (as in our case), this result can be easily proved directly. In fact, if the interior int $\mathcal{M}$ is nonempty, then, taking an arbitrary point $a \in \operatorname{int}\mathcal{M}$, we can define a map ϱ as the projection along the rays emanating from a. In the general case, the construction of ϱ may be realized by using Theorem 3.17 about the relative interior of a convex set in a finite-dimensional space. Do it as an exercise.

Proof of Theorem 26.5. (i) First consider the case when $\mathcal{M}$ is a closed ball B centered at the origin. If a map $f: B \to B$ has no fixed points on the boundary, a map $g: \partial B \times [0, 1] \to \mathbb{R}^n$, $g(x, \lambda) = \lambda f(x)$ generates a homotopy of the vector field $\varphi = i - f$ and the field $\psi(x) \equiv x$. Applying the homotopy invariance and normalization properties we get $\deg(\varphi, \partial B) = \deg(\psi, \partial B) = 1$ and the existence of a fixed point follows from Theorem 26.4.

(ii) In the general case, by using Lemma 26.6, take a closed ball $B \supset \mathcal{M}$ and a retraction $\varrho: B \to \mathcal{M}$. From part (i) it follows that the continuous map $f \circ \varrho: B \to B$ has a fixed point $x_* = f \circ \varrho(x_*)$. But since $x_* \in \operatorname{im}(f) \subset \mathcal{M}$ we have that $\varrho(x_*) = x_*$ and hence x_* is a fixed point of f, concluding the proof. □

5 Heinrich Franz Friedrich Tietze (1880–1964), an Austrian mathematician.

6 James Dugundji (1919–1985), an American mathematician.

Notice that the Brouwer theorem is of a different nature than the contraction map principle. In particular, a fixed point cannot be unique. (For example, if f is the identity map, then each point of $\mathcal{M}$ is a fixed point).

It should be noted that the topological degree can be used not only to provide fixed point results. As an example, we can apply it to prove the following "negative" retraction principle.

Proposition 26.8. *The boundary ∂B of a closed ball $B \subset \mathbb{R}^n$, $n \geq 1$ is not a retract of B.*

Proof. Without loss of generality, we suppose that B is centered at the origin. Assume to the contrary that there exists a retraction $\varrho: B \to \partial B$. Define maps f and g on B by setting $f(x) = x - \varrho(x)$ and $g(x) = 0$. Then f and g coincide on ∂B and hence the corresponding vector fields $\varphi(x) = x - f(x)$ and $\psi(x) \equiv x$ are obviously homotopic. Then $\deg(\varphi, \partial B) = \deg(\psi, \partial B) = 1$. This means that f has a fixed point x_* inside the ball B, hence $\varrho(x_*) = 0$, contrary to the fact that $\varrho(B) = \partial B$. □

It is worth noting that that the above assertion in turn can be used for the proof of fixed and coincidence point results not applying to the retraction principle given in Lemma 26.6. As an example, consider the following statement which also gives us a good opportunity to compare this approach with the topological degree method.

Let $\mathcal{M}$ be a convex compact subset of $\mathbb{R}^n$, $\operatorname{ri}\mathcal{M}$ the relative interior of $\mathcal{M}$, $\operatorname{aff}\mathcal{M}$ its affine hull, and $\partial\mathcal{M} = \mathcal{M} \setminus \operatorname{ri}\mathcal{M}$ the relative boundary of $\mathcal{M}$.

Theorem 26.9. *Suppose that continuous maps $f, g: \mathcal{M} \to \operatorname{aff}\mathcal{M}$ are such that $f(x) \in \mathcal{M}$ for all $x \in \mathcal{M}$ and $g(x) = x$ for all $x \in \partial\mathcal{M}$. Then*

$$\exists x \in \mathcal{M}: f(x) = g(x). \tag{26.1}$$

Proof. First assume that the interior $\operatorname{int}\mathcal{M}$ of the set $\mathcal{M}$ is nonempty. Then $\operatorname{ri}\mathcal{M} = \operatorname{int}\mathcal{M}$ and the relative boundary $\partial\mathcal{M}$ coincides with the boundary of $\mathcal{M}$. We begin with the case when $\mathcal{M}$ is the unit ball B in $\mathbb{R}^n$ centered at the origin with the boundary $\partial\mathcal{M} = S$.

Suppose the contrary, i.e., that $f(x) \neq g(x)$ for all $x \in B$. For two arbitrary different points $x_1 \in B$ and $x_2 \in \mathbb{R}^n$ consider the ray, starting at the point x_1 and passing through the point x_2. Since $x_1 \in B$, this ray intersects the sphere S. Moreover, if $x_1 \notin S$ then the ray intersects S at the unique point which we denote by $R(x_1, x_2)$. In the case $x_1 \in S$, the ray either does not intersect B in points different from x_1, and in this case we set $R(x_1, x_2) = x_1$, or it intersects S at a unique point different from point x_1, which also will be denoted by $R(x_1, x_2)$. (The justification of these facts is a useful exercise.)

It is evident, by construction, that for all $x_1 \in B$, $x_1 \neq x_2$ we have $R(x_1, x_2) \in S$ and, if $x_2 \in S$, then $R(x_1, x_2) = x_2$. In addition, the map R is continuous on the set $(x_1, x_2): x_1 \in B, x_2 \in \mathbb{R}^n, x_1 \neq x_2$.

For each $x \in B$ set $\tilde{R}(x) = R(f(x), g(x))$. Taking into account that by assumption $f(x) \in B$ and $f(x) \neq g(x)$ $\forall x \in B$, the function $\tilde{R}(x)$ is well defined. Also, by construction,

$\tilde{R}(x) \in S\ \forall x \in B$ and $\tilde{R}(x) = x\ \forall x \in S$. Since the map $\tilde{R}$ is obviously continuous, it is the retraction of B on S, contrary to Proposition 26.8.

To demonstrate the other approach, consider a continuous map $\varphi = g - f : \mathcal{M} \to \mathbb{R}^n$. If f has a fixed point on S the assertion is proved. Otherwise, it is easy to see that the map $h: S \times [0, 1] \to \mathbb{R}^n$, $h(x, \lambda) = g(x) - \lambda f(x)$ generates the homotopy of φ and the field $g(x) \equiv x$ on S. This yields $\deg(\varphi, S) = \deg(g, S) = 1$ and hence the map φ has a zero point inside B which is the desired coincidence point of f and g.

The case of an arbitrary convex compact set $\mathcal{M} \subset \mathbb{R}^n$ with a nonempty interior can be reduced to that considered above due to the fact that there exists a homeomorphism $\gamma: B \to \mathcal{M}$. (The construction of such homeomorphism is an easy exercise.)

Suppose now that the interior of $\mathcal{M}$ is empty. However, the relative interior of the convex set $\mathcal{M}$ in $\mathbb{R}^n$ is nonempty (Theorem 3.17). Then, "shifting" the set $\mathcal{M}$ to the origin, i.e., passing from the set $\mathcal{M}$ to the set $\mathcal{M} - x_0$, where x_0 is a certain point from $\mathcal{M}$, and then passing from $\mathbb{R}^n$ to the linear hull $\operatorname{lin}(M - x_0)$, we reduce the general case to the considered one. □

Notice that the Brouwer theorem immediately follows from the above result if we take $g(x) \equiv x$.

Consider the following generalization of Theorem 26.9.

Theorem 26.10. *Let $\mathcal{M}$ be a convex compact set in $\mathbb{R}^n$ and, in addition to continuous maps $f, g: \mathcal{M} \to \operatorname{aff} \mathcal{M}$, for which $f(x) \in \mathcal{M}$ for all $x \in \mathcal{M}$, there exists such a continuous map $\psi: \mathcal{M} \to \mathcal{M}$, that $(g \circ \psi)(x) = x$ for all $x \in \partial\mathcal{M}$. Then* (26.1) *holds true.*

Proof. The validity of this assertion follows from Theorem 26.9, if it were applied to the maps $\tilde{f} = (f \circ \psi)$ and $\tilde{g} = (g \circ \psi)$. We have $f(x) = g(x)$, where $x = \psi(\tilde{x})$, and a point $\tilde{x}$ is a coincidence point of maps $\tilde{f}$ and $\tilde{g}$. □

Passing to continuous vector fields in infinite-dimensional Banach spaces, we discover that in the general case any substantial topological degree theory cannot be suggested for them. This is due to the fundamental fact that in an infinite-dimensional Banach space all continuous nondegenerate vector fields are homotopic to each other (see, e.g., [69]). This makes it possible, for example, to extend every such field from the boundary of an open bounded set to the whole set without zero points, which excludes such general fixed point principles as Theorem 26.4 and so on.

The natural way out of this situation is to narrow the class of fields under consideration. The most simple (and, possibly, most important) class of such fields for which the topological degree theory is well defined can be presented by completely continuous vector fields.

Definition 26.11. Let X be a bounded subset of a Banach space E. We will say that a continuous vector field $\varphi = i - f: X \to E$ is completely continuous if the corresponding map f is compact in the sense that the set $f(X)$ is relatively compact, i.e., the closure $\operatorname{cl} f(X)$ is compact.

Now, let U be a bounded open set in a Banach space E and $\varphi = i - f$ a nondegenerate completely continuous vector field given on the boundary ∂U. We will outline the definition of the topological degree for φ. This characteristic is called *the Leray*[7]*–Schauder*[8] *topological degree* and its construction is based on the following two assertions.

Proposition 26.12. *There exists such an $\alpha > 0$ that*

$$\|\varphi(x)\| > \alpha \quad \textit{for all } x \in \partial U. \tag{26.2}$$

Proof. Supposing the contrary we will have for a certain sequence $\{x_n\} \subset \partial U$:

$$\lim_{n\to\infty} \|x_n - f(x_n)\| = 0.$$

Since the set $f(\partial U)$ is relatively compact we may assume, without loss of generality, that the sequence $\{f(x_n)\}$ converges to a point x_*. But then

$$\lim_{n\to\infty} \|x_n - x_*\| \le \lim_{n\to\infty} \|x_n - f(x_n)\| + \lim_{n\to\infty} \|f(x_n) - x_*\| = 0$$

yielding $x_* \in \partial U$ and $x_* = f(x_*)$, i.e., $\varphi(x_*) = 0$ contradicting to the nondegeneracy of φ. □

Proposition 26.13. *For every $\varepsilon > 0$ there exists a finite-dimensional subspace $E' \subset E$ and a completely continuous vector field $\varphi_\varepsilon = i - f_\varepsilon : \partial U \to E'$ such that*

$$\|f(x) - f_\varepsilon(x)\| < \varepsilon \quad \textit{for all } x \in \partial U.$$

Proof. Choose a finite ε-net $\{x_1, \dots, x_k\}$ of the set $f(\partial U)$ and take its linear hull $\operatorname{lin}\{x_1, \dots, x_k\}$ as the space E'. Consider the operator $P_\varepsilon : E \to E'$ which is called *the Schauder projector* and defined as

$$P\varepsilon(x) = \frac{\sum_{i=1}^k \mu_i(x)}{\sum_{i=1}^k \mu_i(x) x_i},$$

where

$$\mu_i(x) = \max\{0, \varepsilon - \|x - x_i\|\} \quad i = 1, \dots, k.$$

It is an easy exercise to verify that the field $\varphi_\varepsilon = i - f_\varepsilon$, where $f_\varepsilon = P_\varepsilon \circ f$ is the desirable one. □

7 Jean Leray (1906–1998), a French mathematician.

8 Juliusz Schauder (1899–1943), a Polish mathematician.

Notice that, without loss of generality, we may assume in the above proposition that $E' \cap U \neq \emptyset$.

Now, let us take an arbitrary finite-dimensional subspace $E_0 \subset E$ such that $U_0 := U \cap E_0 \neq \emptyset$ and a completely continuous vector field $\varphi_0 = i - f_0$ on ∂U such that $f_0(\partial U) \subset E_0$ and

$$\|f(x) - f_0(x)\| < \frac{\alpha}{2} \quad \text{for all } x \in \partial U,$$

where $\alpha > 0$ is determined by relation (26.2).

Consider on the boundary ∂U_0 the finite-dimensional field $\varphi_0 = i - f_0$. This field is nondegenerate on ∂U_0 since for $x \in \partial U_0$ we have

$$\|\varphi_0(x)\| = \|x - f_0(x)\| \geq \|x - f(x)\| - \|f(x) - f_0(x)\| > \alpha - \frac{\alpha}{2} = \frac{\alpha}{2}.$$

Therefore, for the field φ_0 its topological degree $\deg(\varphi_0, \partial U)$ on ∂U_0 is defined. It can be shown that this characteristic does not depend on the choice of either a finite-dimensional space E_0 or the approximating field φ_0. We call it *the topological degree* $\deg(\varphi, \partial U)$ *of a completely continuous vector field* φ *on* ∂U.

It follows immediately from the definition that the topological degree of a completely continuous vector field in a Banach space has the same main properties (i)–(iii) as the degree of a finite-dimensional field. (It is necessary to remember only that the homotopy of completely continuous fields in property (ii) should be generated by the compact map $g: \partial U \times [0, 1] \to E$.)

The analog of the nonzero degree principle (Theorem 26.4) also holds true. Among its numerous consequences, select the following *Schauder fixed point theorem* which is the direct generalization of the Brouwer theorem.

Theorem 26.14. *Let $\mathcal{M}$ be a nonempty convex closed bounded subset of a Banach space E. Then each continuous compact map $f: \mathcal{M} \to \mathcal{M}$ has a fixed point.*

26.2 Topological degree of set-valued maps

In this section we will extend the topological degree theory to set-valued maps in Banach spaces.

Let, as earlier, E be a Banach space, by the symbol $\mathcal{H}_{cv}(E)$ we denote the collection of all nonempty compact convex subsets of E. For $X \subset E$, in accordance with the terminology used in the previous section, we say that a set-valued map $F: X \to \mathcal{H}_{cv}(E)$ defines *a set-valued vector field* $\Phi = i - F: X \to \mathcal{H}_{cv}(E)$, $\Phi(x) = x - F(x)$. A point $x \in X$ such that $0 \in \Phi(x)$ is called *a zero point of a field* Φ. It is clear that zero points of a set-valued field $\Phi = i - F$ coincide with fixed points of a map F. If a set-valued vector field Φ has no zero points, we will call it *nondegenerate*.

Given a bounded set $X \subset E$, we say that an upper semicontinuous set-valued map $F\colon X \to \mathcal{H}_{cv}(E)$ is *completely upper semicontinuous* (in short: *completely u.s.c.*) if the set $F(X)$ is relatively compact. For convenience, the corresponding field $\Phi = i - F$ we will also call completely u.s.c.

Recall that by the Mazur theorem, the closure of a convex hull of a relatively compact subset of a Banach space is compact (see Chapter 2). In particular, from here it follows that if $F_0, F_1\colon X \to \mathcal{H}_{cv}(E)$ are completely u.s.c. set-valued maps, then the set-valued map $G\colon X \times [0, 1] \to \mathcal{H}_{cv}(E)$, given as $G(x, \lambda) = \lambda F_1(x) + (1 - \lambda)F_0(x)$ is also completely u.s.c.

Now, let U be an open bounded subset of E.

Definition 26.15. Completely u.s.c. set-valued maps $F_0, F_1\colon \partial U \to \mathcal{H}_{cv}(E)$ and the corresponding set-valued fields $\Phi_0 = i - F_0$, $\Phi_1 = i - F_1$ are called homotopic,

$$\Phi_0 \sim \Phi_1$$

if there exists a completely u.s.c. set-valued map $G\colon \partial U \times [0, 1] \to \mathcal{H}_{cv}(E)$ such that
(1) $x \notin G(x, \lambda)$ for all $x \in \partial U, \lambda \in [0, 1]$;
(2) $G(\cdot, 0) = F_0, G(\cdot, 1) = F_1$.

It is easy to verify that the relation of homotopy is the equivalence relation on the class of all nondegenerate completely u.s.c. set-valued vector fields defined on ∂U.

Definition 26.16. A completely continuous single-valued vector field $\varphi = i - f, f\colon \partial U \to E$ is called a single-valued homotopic approximation of a completely u.s.c. set-valued vector field $\Phi = i - F, F\colon \partial U \to \mathcal{H}_{cv}(E)$ if $\varphi \sim \Phi$.

To prove the existence of a single-valued homotopic approximation we need the following assertion.

Lemma 26.17. *Let $X \subset E$ be a bounded closed set; Δ a compact metric space, $F\colon X \times \Delta \to \mathcal{H}_{cv}(E)$ a completely u.s.c. set-valued map such that $F(\cdot, \mu)$ is fixed point free on a closed subset $X_1 \subset X$ for all $\mu \in \Delta$. Then for all sufficiently small $\varepsilon > 0$, if $f_\varepsilon\colon X \times \Delta \to E$ is a regular ε-approximation of F, then the set-valued map $G\colon X \times \Delta \times [0, 1] \to \mathcal{H}_{cv}(E)$,*

$$G(x, \mu, \lambda) = \lambda f_\varepsilon(x, \mu) + (1 - \lambda)F(x, \mu)$$

is such that $x \notin G(x, \mu, \lambda)$ for all $(x, \mu, \lambda) \in X_1 \times \Delta \times [0, 1]$.

Proof. Supposing the contrary, we will have sequences $\{\varepsilon_n\}_{n=1}^\infty$, $\varepsilon_n > 0$, $\varepsilon_n \to 0$; $\{x_n\}_{n=1}^\infty \subset X_1, \{\mu_n\}_{n=1}^\infty \subset \Delta$ and $\{\lambda_n\}_{n=1}^\infty \subset [0, 1]$ such that

$$x_n \in \lambda_n f_{\varepsilon_n}(x_n, \mu_n) + (1 - \lambda_n)F(x_n, \mu_n)$$

for each $n = 1, 2, \ldots,$ where $f_{\varepsilon_n}\colon X \times \Delta \to E$ are regular ε_n-approximations of F. The sequences $\{f_{\varepsilon_n}(x_n, \mu_n)\}$ and $\{x_n\}$ are contained in a compact set $\mathrm{cl}\,\mathrm{conv}\,F(X)$ and hence we may assume, without loss of generality, that they are convergent: $f_{\varepsilon_n}(x_n, \mu_n) \to v_0$, $x_n \to x_0 \in X_1$. It is clear that we may assume that the sequences $\{\mu_n\}$ and $\{\lambda_n\}$ are also convergent: $\mu_n \to \mu_0 \in \Delta$, $\lambda_n \to \lambda_0 \in [0, 1]$.

For points x_n we have the representation

$$x_n = \lambda_n f_{\varepsilon_n}(x_n, \mu_n) + (1 - \lambda_n) y_n,$$

where $y_n \in F(x_n, \mu_n)$. Since F is completely u.s.c., we also assume that $y_n \to y_0$, and, by the closedness of F we get $y_0 \in F(x_0, \mu_0)$. Each point of the form $((x_n, \mu_n), f_{\varepsilon_n}(x_n, \mu_n)) \in X_1 \times \Delta \times E$ lies in the ε_n-neighborhood of the graph of F, hence the limit point (x_0, μ_0, v_0) belongs to this graph, i.e., $v_0 \in F(x_0, \mu_0)$. Passing to the limit, we get

$$x_0 = \lambda_0 v_0 + (1 - \lambda_0) y_0 \in F(x_0, \mu_0),$$

that contradicts the absence of fixed points of $F(\cdot, \mu_0)$ on X_1. □

Proposition 26.18. *Each nondegenerate completely u.s.c. set-valued vector field $\Phi = i - F$, $F: \partial U \to \mathcal{K}_{cv}(E)$ admits a single-valued homotopic approximation.*

Proof. From the previous lemma it follows that as such an approximation we may take the field $\varphi = i - f$, where $f: \partial U \to E$ is an arbitrary regular single-valued ε-approximation of F and $\varepsilon > 0$ is sufficiently small. □

We can now give the following definition.

Definition 26.19. The topological degree

$$\deg(\Phi, \partial U)$$

of a nondegenerate completely u.s.c. set-valued vector field $\Phi = i - F$, $F: \partial U \to \mathcal{K}_{cv}(E)$ is the topological degree $\deg(\varphi, \partial U)$ of its arbitrary single-valued homotopic approximation φ.

Let us show that this characteristic is well defined, i.e., the topological degree $\deg(\Phi, \partial U)$ does not depend on the choice of a single-valued homotopic approximation.

Lemma 26.20. *Let $\varphi_0 = i - f_0$ and $\varphi_1 = i - f_1$ be single-valued homotopic approximations of a nondegenerate completely u.s.c. set-valued vector field $\Phi = i - F$, $F: \partial U \to \mathcal{K}_{cv}(E)$. Then φ_0 and φ_1 are homotopic in the class of single-valued completely continuous vector fields.*

Proof. From Definition 26.16 it follows that the fields φ_0 and φ_1 are homotopic in the class of completely u.s.c. set-valued fields. Let this homotopy be generated by a fixed point free completely u.s.c. set-valued map $G: \partial U \times [0, 1] \to \mathcal{K}_{cv}(E)$, $G(\cdot, 0) = f_0$, $G(\cdot, 1) = f_1$.

According to Lemma 26.17 there exists a regular single-valued ε-approximation $g\colon \partial U \times [0,1] \to E$ of the set-valued map G such that

$$x \notin \lambda g(x,\mu) + (1-\lambda)G(x,\mu)$$

for all $x \in \partial U$, $\mu \in [0,1]$, $\lambda \in [0,1]$.

Then it is easy to see that the desired homotopy connecting φ_0 and φ_1 is generated by a completely continuous map $h\colon \partial U \times [0,1] \to E$,

$$h(x,\nu) = \begin{cases} 3\nu g(x,0) + (1-3\nu)f_0(x), & 0 \le \nu \le \frac{1}{3}; \\ g(x, 3\nu - 1), & \frac{1}{3} \le \nu \le \frac{2}{3}; \\ (3-3\nu)g(x,1) + (3\nu-2)f_1(x), & \frac{2}{3} \le \nu \le 1, \end{cases}$$ □

From the proved lemma and the property of homotopy invariance of topological degree of single-valued completely continuous vector fields it follows that degrees of all single-valued homotopic approximations of the set-valued field Φ are the same, which demonstrates the correctness of Definition 26.19.

Consider now the principal properties of the defined characteristic.

Theorem 26.21 (Normalization property). *If $F(x) \equiv A$ for all $x \in \partial U$, where $A \subset E$ is a compact convex subset, then*

$$\deg(i - F, \partial U) = \begin{cases} 1, & \text{if } A \subset U, \\ 0, & \text{if } A \cap \operatorname{cl} U = \emptyset. \end{cases}$$

Proof. It is sufficient to take as a single-valued homotopic approximation a field of the form $\varphi = i - f$, where $f(x) \equiv x_0 \in A$. □

Theorem 26.22 (Homotopy invariance property). *If $\Phi_0 \sim \Phi_1$, then*

$$\deg(\Phi_0, \partial U) = \deg(\Phi_1, \partial U).$$

Proof. This assertion follows immediately from Definition 26.19 and the transitivity property of the homotopy relation. □

Notice that concrete conditions of homotopy are similar to those for single-valued fields. For example, the following analogue of the Poincare–Bohl theorem holds true.

Theorem 26.23. *Suppose that nondegenerate completely u.s.c. set-valued vector fields $\Phi_0 = i - F_0$, $\Phi_1 = i - F_1$ do not admit opposite directions, i.e.,*

$$\frac{z_0}{\|z_0\|} \neq -\frac{z_1}{\|z_1\|}$$

for each $z_0 \in \Phi_0(x)$, $z_1 \in \Phi_1(x)$, $x \in \partial U$. Then $\Phi_0 \sim \Phi_1$ and hence

$$\deg(\Phi_0, \partial U) = \deg(\Phi_1, \partial U).$$

Proof. The homotopy of set-valued fields Φ_0 and Φ_1 is generated by the completely u.s.c. set-valued map $G\colon \partial U \times [0,1] \to \mathcal{H}_{cv}(E)$, $G(x,\lambda) = \lambda F_1(x) + (1-\lambda)F_0(x)$. Indeed, if $x_0 \in G(x_0, \lambda_0)$ for certain $x_0 \in \partial U$ and $\lambda_0 \in (0,1)$, then $x_0 = \lambda_0 y_1 + (1-\lambda_0)y_0$, where $y_1 \in F_1(x_0)$, $y_0 \in F_0(x_0)$. But it is equivalent to

$$x_0 - y_0 = -\frac{\lambda_0}{1-\lambda_0}(x_0 - y_1),$$

contrary to the condition of the theorem. □

Consider now the following property of the topological degree.

Theorem 26.24 (Additive dependence on the domain property). *Let $\{U_j\}_{j\in J}$ be a family disjoint open subsets of U; a set-valued map $F\colon \operatorname{cl} U \to \mathcal{H}_{cv}(E)$ is completely u.s.c. and fixed point free on the set $\operatorname{cl} U \setminus \bigcup_{j\in J} U_j$. Then the topological degrees $\deg(i-F, \partial U_j)$ are nonzero for only finite many indices $j \in J$ and*

$$\deg(i-F, \partial U) = \sum_{j\in J} \deg(i-F, \partial U_j).$$

Proof. Let $f\colon \operatorname{cl} U \to E$ be a regular ε-approximation of the set-valued map F. From Lemma 26.17 it follows that if $\varepsilon > 0$ is sufficiently small, then f is fixed point free on $\operatorname{cl} U \setminus \bigcup_{j\in J} U_j$ and the restriction of f on each boundary ∂U, ∂U_j, $j \in J$ is a single-valued homotopic approximation of the restriction of F to the corresponding boundary. Then the assertion follows from the analogous principle for single-valued fields. □

Let us present now the main fixed point principle.

Theorem 26.25. *Let a completely u.s.c. set-valued map $F\colon \operatorname{cl} U \to \mathcal{H}_{cv}(E)$ be fixed point free on ∂U and*

$$\deg(i-F, \partial U) \neq 0.$$

Then F has a fixed point in U.

Proof. If we suppose, to the contrary, that the set-valued map F is fixed point free on the whole $\operatorname{cl} U$ then, applying Lemma 26.17, we construct for a sufficiently small $\varepsilon > 0$ such a regular single-valued ε-approximation $f\colon \operatorname{cl} U \to E$ of F which is also fixed point free and its restriction to the boundary ∂U is a single-valued homotopic approximation of the restriction of F to the same boundary. But then

$$\deg(i-F, \partial U) = \deg(i-f, \partial U) = 0,$$

which contradicts the assumption. □

This general principle also yields a number of fixed point results for set-valued maps. For example, by using the same reasoning as during the proof of Theorem 26.5, we get the following *Bohnenblust*[9]*–Karlin*[10] *fixed point theorem* [73], which is the set-valued analog of the Schauder fixed point theorem.

Theorem 26.26. *Let $\mathcal{M}$ be a nonempty convex closed bounded subset of a Banach space E. Then each completely u.s.c. set-valued map $F: \mathcal{M} \to \mathcal{H}_{cv}(\mathcal{M})$ has a fixed point.*

For a finite-dimensional space E this assertion, which is the set-valued generalization of the Brouwer fixed point theorem, is called *the Kakutani*[11] *fixed point theorem* [74].

Another example is presented by the following result.

Theorem 26.27. *Let $U \subset E$ be an open bounded neighborhood of zero, $F: \operatorname{cl} U \to \mathcal{H}_{cv}(E)$ a completely u.s.c. set-valued map such that*

$$x \notin \lambda F(x) \quad \textit{for all } x \in \partial U \textit{ and } 0 < \lambda \leq 1.$$

Then F has a fixed point in U.

Proof. The completely u.s.c. set-valued map $G: \partial U \times [0,1] \to \mathcal{H}_{cv}(E)$, $G(x,\lambda) = \lambda F(x)$ generates the homotopy of the set-valued field $\Phi = i - F$ and the field $\psi(x) \equiv x$. Therefore

$$\deg(\Phi, \partial U) = \deg(\psi, \partial U) = 1,$$

and we can apply Theorem 26.25. □

9 Henri Frederic Bohnenblust (1906–2000), a Swiss and American mathematician.
10 Samuel Karlin (1924–2007), an American mathematician.
11 Shizuo Kakutani (1911–2004), a Japanese and American mathematician.

27 Existence results for differential inclusions via the fixed point method

In this section we will apply the topological degree theory developed in the previous section to the existence of solutions to the Cauchy problem for a differential inclusion of the following form

$$x'(t) \in F(t, x(t)) \tag{27.1}$$

$$x(t_0) = x_0, \tag{27.2}$$

where $F: I \times \mathbb{R}^n \to \mathcal{H}_{cv}(\mathbb{R}^n)$ is an L^1-upper Carathéodory set-valued map (see Definition 20.8), $I = [t_0, T]$ is a certain interval endowed with the Lebesgue measure.

As in Section 2.10, by *a solution of problem* (27.1), (27.2) *on an interval* $t_0, \tau)$ with $t_0 < \tau \le T$ we mean an absolutely continuous function $x: [t_0, \tau] \to \mathbb{R}^n$ satisfying initial condition (27.2) and inclusion (27.1) in almost each point of the interval $[t_0, \tau]$.

Our methods in the search for solutions to problem (27.1), (27.2) are based on the following assertion, whose proof is an easy exercise. Consider the integral set-valued operator

$$j \circ \mathcal{P}_F: C([t_0, \tau]; \mathbb{R}^n) \to \mathcal{H}_{clv}(C([t_0, \tau]; \mathbb{R}^n))$$

which is defined for every $\tau \in (t_0, T]$ (see Chapter 23).

Proposition 27.1. *A function* $x: [t_0, \tau] \to \mathbb{R}^n$ *is a solution of problem* (27.1), (27.2) *if and only if it is a fixed point of the set-valued operator*

$$\mathcal{J}_F: C([t_0, \tau]; \mathbb{R}^n) \to \mathcal{H}_{clv}(C([t_0, \tau]; \mathbb{R}^n)),$$
$$\mathcal{J}_F = x_0 + j \circ \mathcal{P}_F.$$

In the following we will need the following property of the set-valued operator$\mathcal{J}_F$.

Lemma 27.2. *The image* $\mathcal{J}_F(D) \subset C([t_0, \tau]; \mathbb{R}^n)$ *of each bounded set* $D \subset C([t_0, \tau]; \mathbb{R}^n)$ *is relatively compact.*

Proof. We will use the criterion of compactness given by the Arcela–Ascoli theorem. Let $y \in \mathcal{J}_F(D)$ be an arbitrary function. This means that y has the form

$$y(t) = x_0 + \int_{t_0}^{t} f(s)\, ds,$$

where $f \in \mathcal{P}_F(x)$, $x \in D$. It is clear that the set $\Omega \subset \mathbb{R}^n$ which is the union of ranges of functions from D is bounded. Then for each $t \in [t_0, \tau]$ we have

$$\|y(t)\| \le \|x_0\| + \int_{t_0}^{t} \|f(s)\|\, ds \le \|x_0\| + \int_{t_0}^{\tau} \nu_\Omega(s)\, ds,$$

i.e., the set $\mathcal{J}_F(D)$ is bounded.

DOI 10.1515/9783110460308-028

From the other side, for each $t_1, t_2 \in [t_0, \tau]$ we have

$$\|y(t_2) - y(t_1)\| = \|\int_{t_1}^{t_2} f(s)\,ds\| \le \int_{t_1}^{t_2} \|f(s)\|\,ds \le \int_{t_1}^{t_2} v_\Omega(s)\,ds,$$

i.e., the set $\mathfrak{J}_F(D)$ is equicontinuous. □

We may prove now the following *local existence theorem.*

Theorem 27.3. *Let a set-valued map $F\colon I \times \mathbb{R}^n \to \mathcal{H}_{cv}(\mathbb{R}^n)$ satisfy L^1-upper Carathéodory conditions. Then there exists such a τ, $t_0 < \tau \le T$, that Cauchy problem* (27.1), (27.2) *has a solution on the interval $[t_0, \tau]$.*

Proof. Take an arbitrary $R > 0$ and, for the number $r = \|x_0\| + R$, let v_r be the corresponding, by condition $(F3)$, function. Let now τ, $t_0 < \tau \le T$ be sufficiently close to t_0 to provide

$$\int_{t_0}^{\tau} v_r(s)\,ds \le R.$$

Then for the ball $\mathcal{B} \subset C([t_0, \tau]; \mathbb{R}^n)$,

$$\mathcal{B} = \{y \in C([t_0, \tau]; \mathbb{R}^n) : \|y - \overline{x}_0\|_C \le R\},$$

where $\overline{x}_0(t) \equiv x_0$, we obviously will have

$$\mathfrak{J}_F(\mathcal{B}) \subset \mathcal{B}.$$

From Corollary 20.14 easily follows the closedness of the set-valued map $\mathfrak{J}_F : \mathcal{B} \to \mathcal{H}_{clv}(\mathcal{B})$, and from Lemma 27.2 and Theorem 17.15 we conclude that it is upper u.s.c. Applying the Bohnenblust–Karlin theorem (Theorem 26.26) we obtain that the set-valued map $\mathfrak{J}_F$ has a fixed point that, by Proposition 27.1 completes the proof. □

In the following we need the following slightly modified assertion about integral inequalities known under the title of *the Gronwall*[1] *lemma* (see, for example, [75], Theorem III.1.1).

Lemma 27.4. *Let $u, v\colon [a, b] \to \mathbb{R}$ be nonnegative functions, u integrable and v continuous; $C \ge 0$ a certain constant and*

$$v(t) \le C + \int_a^t u(s)v(s)\,ds, \quad a \le t \le b.$$

Then

$$v(t) \le C \exp \int_a^t u(s)\,ds, \quad a \le t \le b.$$

1 Thomas Hakon Grönwall (1877–1932), a Swedish mathematician.

Replace now condition (iii) of local integral boundedness of the right-hand part F of the inclusion with the stronger condition of *sublinear growth*

(iii′) there exists such a function $\alpha \in L^1_+(I)$, that

$$\|F(t, x)\| \leq \alpha(t)(1 + \|x\|)$$

for all $x \in \mathbb{R}^n$ and a.e. $t \in I$.

Now, we can present *the global existence result.*

Theorem 27.5. *Let a set-valued map $F\colon I \times \mathbb{R}^n \to \mathcal{H}_{cv}(\mathbb{R}^n)$ satisfy the upper Carathéodory conditions and the condition of sublinear growth* (iii′)*. Then Cauchy problem* (27.1), (27.2) *has a solution on the interval* $[t_0, T]$.

For the proof we need the following assertion.

Lemma 27.6. *Under conditions of Theorem 27.5 the set of all solutions to a one-parameter family of the Cauchy problems*

$$x'(t) \in \lambda F(t, x(t)), \quad t \in I, \ \lambda \in [0, 1] \tag{27.3}$$

$$x(t_0) = x_0, \tag{27.4}$$

is a priori bounded.

Proof. Each function that is a solution of the above problem for a certain $\lambda \in [0, 1]$ has the form

$$x(t) = x_0 + \lambda \int_{t_0}^{t} f(s)\, ds,$$

where $f \in \mathcal{P}_F(x)$. But then for a continuous function $v(t) = \|x(t)\|$ we have the following estimate:

$$\begin{aligned} v(t) \leq \|x_0\| + \lambda \int_{t_0}^{t} \|f(s)\|\, ds &\leq \|x_0\| + \int_{t_0}^{t} \alpha(s)(1 + \|x(s)\|)\, ds \\ &\leq \|x_0\| + \int_{t_0}^{T} \alpha(s)\, ds + \int_{t_0}^{t} \alpha(s) v(s)\, ds. \end{aligned}$$

Applying Lemma 27.4 we get

$$\|x(t)\| \leq C \exp \int_{t_0}^{T} \alpha(s)\, ds, \quad t \in [t_0, T],$$

where $C = \|x_0\| + \int_{t_0}^{T} \alpha(s)\, ds$. □

Proof of Theorem 27.5. For simplicity, without loss of generality, we will assume that $x_0 = 0$. By virtue of Lemma 27.6 there exists an open ball $B \subset C(I; \mathbb{R}^n)$ centered at zero containing all solutions of problem (27.3), (27.4).

It is easy to see that the set-valued map

$$\mathcal{G}_F \colon \operatorname{cl} B \times [0, 1] \to \mathcal{H}_{\mathrm{cv}}(C(I; \mathbb{R}^n))$$
$$\mathcal{G}_F(x, \lambda) = \lambda j \circ \mathcal{P}_F(x)$$

is completely u.s.c. and, moreover, $x \notin \mathcal{G}_F(x, \lambda)$ for all $(x, \lambda) \in \partial B \times [0, 1]$, since otherwise such a function x should be a solution of problem (27.3) (27.4). The proof is completed by the application of Theorem 26.27. ☐

Notation

$[x_1, x_2]$ closed interval, connecting points x_1 and x_2

$[x_1, x_2), (x_1, x_2]$ semi-intervals

(x_1, x_2) interval

$\operatorname{conv} A$ convex hull of the set A

$\operatorname{lin} A$ linear hull of the set A

$\operatorname{aff} A$ affine hull of the set A

$\dim X$ dimension of the linear space X

$\operatorname{ext}(A)$ set of extreme points of the set A

$\operatorname{cl} A$ closure of the set A

∂A boundary of the set A

$\operatorname{int} A$ interior of the set A – set of all interior points of the set A

$\operatorname{ri} A$ relative interior of the convex set A

$\|x\|$ norm of the element x

$\|A\|$ norm of the set A, i.e., the supremum of the norms of elements of A

$|x| = (\langle x, x\rangle)^{1/2}$ module of the vector x, i.e., the Euclidean norm of the element x

X^* (topologically) conjugate space to the normed space X, i.e., the space of linear continuous functionals on X

$A + B = \{x\colon x = x_1 + x_2,\ x_1 \in A, x_2 \in B\}$ sum of sets in the Minkowski sense

$\alpha A = \{x\colon x = \alpha x_1,\ x_1 \in A\}$ product of the set A by the number α

$K^* = \{x^* \in X\colon \langle x^*, x\rangle \geq 0\ \forall x \in K\}$ conjugate cone to K; $K^0 = \{x^* \in X\colon \langle x^*, x\rangle \leq 0,\ \forall x \in K\}$ – normal cone to the cone K

$K^{**} = (K^*)^*$ second conjugate to the cone K

$O(x_0, \varepsilon) = \{x \in X\colon \|x - x_0\| < \varepsilon\}$ open ε-neighborhood of the point x_0 in the normed space X

$O_\varepsilon = O(0, \varepsilon)$ open ε-neighborhood of the point 0 in a normed space

$A_\varepsilon = A + O_\varepsilon$ open ε-neighborhood of the set A in a normed space

(X, τ) topological space, i.e., the set X with the topology τ defined on it

ρ metric

$\rho(x_1, x_2)$ distance between the elements x_1 and x_2

(X, ρ) metric space, i.e., the set X with the metric ρ defined on it

$O^X(x_0, \varepsilon) = \{x \in X\colon \rho_X(x, x_0) < \varepsilon\}$ open ε-neighborhood of the point x_0 in the metric space (X, ρ)

$B^X(x_0, \varepsilon) = \{x \in X\colon \rho_X(x, x_0) \leq \varepsilon\}$ closed ε-neighborhood of the point x_0 in the metric space (X, ρ)

$X \times Y$ Cartesian product of the metric spaces (X, ρ_X) and (Y, ρ_Y), consisting of ordered pairs (x, y), $x \in X, y \in Y$ with the metric $\rho((x_1, y_1), (x_2, y_2)) = \rho_X(x_1, x_2) + \rho_Y(y_1, y_2)$

$h(M, N) = \inf\{r > 0\colon O^X(M, r) \supseteq N,\ O^X(N, r) \supseteq M\}$ Hausdorff distance between the sets M and N; $\operatorname{dist}(M, N) = \inf\{\rho(x, y),\ x \in M,\ y \in N\}$ – distance between the sets M and N

$h^+(M, N) = \inf\{\varepsilon > 0\colon O^X(N, \varepsilon) \supset M\}$ deviation of the set M from the set N

$O^X(A, \varepsilon) = \{x \in X\colon \operatorname{dist}(x, A) < \varepsilon\}$ open ε-neighborhood of the set A in the metric space (X, ρ)

$B^X(A, \varepsilon) = \{x \in X\colon \operatorname{dist}(x, A) \leq \varepsilon\}$ closed ε-neighborhood of the set A in the metric space (X, ρ)

DOI 10.1515/9783110460308-029

$\mathbb{N}$ set of natural numbers

$\mathbb{R}$ set of real numbers

$\bar{\mathbb{R}} = \mathbb{R} \cup \{-\infty, +\infty\}$ extended set of real numbers

$\mathbb{R}^n$ n-dimensional arithmetic space

$C[a, b]$ linear space of functions, continuous on the interval $[a, b]$

$C(T)$ normed space of continuous bounded functions $f: T \to \mathbb{R}$ with the norm $\|f\| = \sup\{|f(t)|: t \in T\}$

$C(I; E)$ Banach space of continuous functions $f: I \to E$ defined on the closed bounded interval $I \subset \mathbb{R}$ with the values in the Banach space E with the norm $\|f\| = \sup\{\|f(t)\|_E: t \in I\}$

$L^1(I; E)$ Banach space of Bochner integrable functions $f: I \to E$ defined on the closed bounded interval $I \subset \mathbb{R}$ with the values in the Banach space E with the norm $\|f\| = \int_I \{\|f(t)\|_E\, dt\}$

l_p normed space of sequences $x = (x^1, x^2, \dots)$, for which the series $\sum_{i=1}^{\infty} |x^i|^p$ converges, endowed with the norm $\|x\| = (\sum_{i=1}^{\infty} |x^i|^p)^{1/p}$

l_∞ normed space of bounded sequences $x = (x^1, x^2, \dots)$ endowed with the norm $\|x\| = \sup_i \{|x^i|\}$

$\operatorname{epi} f = \{(x, \alpha) \in X \times \mathbb{R}: f(x) \leq \alpha\}$ epigraph of the function f

$\operatorname{dom} f = \{x \in X: f(x) < +\infty\}$ domain of the function f

$\operatorname{gph} F = \{(x, y) \in X \times Y: y \in F(x)\}$ graph of the set-valued map F

$f_1 \oplus \dots \oplus f_m$ infimal convolution of the functions $f_1, f_2, \dots, f_m$

$\operatorname{cl} f$ closure of the function f

$\mathcal{L}_a f = \{x \in X: f(x) \leq a\}$ Lebesgue set of the function f

$f|_C$ restriction of the map $f: X \to Y$ to the set $C \subset X$

f^* function conjugate to the function f

$f^{**} = (f^*)^*$ second conjugate to the function f

$c(x^*, A) = \sup_{y \in A} \langle x^*, y \rangle$ support function of the set A

δ_A indicator function of the set A

$P_A(x) = \inf\{r > 0: x/r \in A\}$ Minkowski functional of the set A

$f'(x; y)$ derivative of the function f at the point $x \in \mathbb{R}^n$ along the direction $y \in \mathbb{R}^n$

$\partial f(x)$ subdifferential of the convex function f at the point x

∂f subdifferential map

$\langle x^*, x \rangle$ action of the linear functional x^* on the vector $x \in X$

$\mathcal{H}(X)$ collection of all nonempty closed bounded subsets of the metric space (X, ρ)

$\mathcal{H}_c(X)$ collection of all nonempty compact subsets of the metric space (X, ρ)

$\mathcal{H}_{cv}(X)$ collection of all nonempty compact convex subsets of the normed space X

$\mathcal{H}_{clv}(X)$ collection of all nonempty closed convex subsets of the normed space X
$\mathcal{F}(G) = \{K \in \mathcal{H}_c(X): K \bigcap G \neq \emptyset\}$; $\mathcal{L}(G) = \{K \in \mathcal{H}_c(X): K \subset G\}$

$(\bigcup_{j \in J} F_j)(x) = \bigcup_{j \in J} F_j(x)$ union of the set-valued maps F_j

$(\bigcap_{j \in J} F_j)(x) = \bigcap_{j \in J} F_j(x)$ intersection of the set-valued maps F_j

$F^-(B) = \{x \in X: F(x) \bigcap B \neq \emptyset\}$ complete pre-image of the set $B \subset Y$

$\mathcal{P}_F$ superposition set-valued operator generated by the set-valued map F

TM_x tangential space to the manifold M at the point x

S_φ set of all singular points of the map φ

$\varphi_0 \sim \varphi_1$, $\Phi_0 \sim \Phi_1$ homotopy of the single-valued (respectively, the set-valued) vector fields

$\deg \varphi$ topological degree of the map φ

$\deg(\varphi, \partial U)$, $\deg(\Phi, \partial U)$ topological degree of the single-valued vector field φ (respectively, the set-valued vector field Φ) on the boundary of the set U

Bibliography

[1] Arutyunov AV. Lectures on Convex and Set-Valued Analysis. Moscow: Fizmatlit; 2014.
[2] Rockafellar RT. Convex Analysis. Princeton, NJ: Princeton University Press; 1970. (Princeton Mathematical Series; vol. 28).
[3] Alekseev VM, Tikhomirov VM, Fomin SV. Optimal Control. New York: Consultants Bureau; 1987. (Contemporary Soviet Mathematics).
[4] Borwein JM, Lewis AS. Convex Analysis and Nonlinear Optimization. Theory and Examples, 2nd ed. New York: Springer; 2006. (CMS Books in Mathematics; vol. 3).
[5] Castaing C, Valadier M. Convex Analysis and Measurable Multifunctions. Berlin, Heidelberg, New York: Springer-Verlag; 1977. (Lecture Notes in Math.; vol. 580).
[6] Ekeland I, Temam R. Convex Analysis and Variational Problems. Amsterdam: North Holland; 1976.
[7] Ioffe AD, Tikhomirov VM. Theory of Extremal Problems. Amsterdam: North Holland; 1979.
[8] Magaril-Il'yaev GG, Tikhomirov VM. Convex Analysis: Theory and Applications. Providence, RI: American Mathematical Society; 2003. (Translations of Mathematical Monographs; vol. 222).
[9] Mordukhovich BS. Variational Analysis and Generalized Differentiation. I. Basic theory. Berlin: Springer-Verlag; 2006. (Grundlehren der Mathematischen Wissenschaften; vol. 330).
[10] Mordukhovich BS, Nam NM. An Easy Path to Convex Analysis and Applications. St. Rafael, CA: Morgan and Claypool Publishers; 2014.
[11] Odinets VP, Slezak VA. Foundations of Convex Analysis, NIC. Izhevsk: Regular and Chaotic Dynamics; 2010.
[12] Pshenichnyi BN. Convex Analysis and Extremal Problems. Moscow: Nauka; 1980.
[13] Polovinkin ES, Balashov MV. Elements of Convex and Strongly Convex Analysis. Moscow: Fizmatlit; 2004.
[14] Rockafellar RT, Wets RJB. Variational Analysis. Berlin: Springer-Verlag; 1998. (Grundlehren der Mathematischen Wissenschaften; vol. 317).
[15] Aubin JP, Cellina A. Differential Inclusions. Set-valued Maps and Viability Theory. Berlin, Heidelberg, New York, Tokyo: Springer-Verlag; 1984.
[16] Aubin JP, Frankowska H. Set-Valued Analysis. Boston, Basel, Berlin: Birkhäuser; 1990.
[17] Borisovich YG, Gel'man BD, Myshkis AD, Obukhovskii VV. Introduction to the Theory of Multi-valued Mappings and Differential Inclusions, 2nd ed. Moscow: Knizhnyi Dom "LIBROKOM"; 2011.
[18] Deimling K. Nonlinear Functional Analysis. Berlin: Springer-Verlag; 1985.
[19] Deimling K. Multivalued Differential Equations. Berlin: Walter de Gruyter; 1992. (De Gruyter Series in Nonlinear Analysis and Applications; vol. 1).
[20] Górniewicz L. Topological Fixed Point Theory of Multivalued Mappings, 2nd ed. Dordrecht: Springer; 2006. (Topological Fixed Point Theory and Its Applications; vol. 4).
[21] Hu S, Papageorgiou NS. Handbook of Multivalued Analysis. Vol. I. Theory. Dordrecht: Kluwer Academic Publishers; 1997. (Mathematics and its Applications; vol. 419).
[22] Kamenskii M, Obukhovskii V, Zecca P. Condensing Multivalued Maps and Semilinear Differential Inclusions in Banach Spaces. Berlin, New York: Walter de Gruyter; 2001. (De Gruyter Series in Nonlinear Analysis and Applications; vol. 7).
[23] Polovinkin ES. Set-Valued Analysis and Differential; Inclusions. Moscow: Fizmatlit; 2014.
[24] Warga J. Optimal Control of Differential and Functional Equations. New York, London: Academic Press; 1972.
[25] Dieudonné J. Foundations of Modern Analysis. New York, London: Academic Press; 1960.
[26] Dunford N, Schwartz JT. Linear Operators. Part I. General Theory. New York: Interscience; 1958.

DOI 10.1515/9783110460308-030

[27] Kolmogorov AN, Fomin SV. Elements of the Theory of Functions and Functional Analysis. Mineola, NY: Dover Publications, Inc.; 1999.
[28] Kolmogorov AN, Fomin SV. Introductory Real Analysis. New York: Dover Publications, Inc.; 1975.
[29] Ljusternik LA, Sobolev VI. Elemente der Funktionalanalysis. Berlin: Akademie-Verlag; 1979. (Mathematische Lehrbücher und Monographien, I. Abteilung: Mathematische Lehrbücher; vol. 8).
[30] Rudin W. Functional Analysis, 2nd ed. New York: McGraw-Hill, Inc.; 1991. (International Series in Pure and Applied Mathematics).
[31] Schwartz L. Analyse Mathematique. Paris: Hermann; 1967.
[32] Hardy GH, Littlewood JE, Pólya G. Inequalities. Cambridge: Cambridge University Press; 1952.
[33] Clarke FH. Optimization and Nonsmooth Analysis, 2nd ed. Philadelphia, PA: Society for Industrial and Applied Mathematics (SIAM); 1990. (Classics in Applied Mathematics; vol. 5).
[34] Evans LC, Gariepy RF. Measure Theory and Fine Properties of Functions, revised ed. Boca Raton, FL: CRC Press; 2015. (Textbooks in Mathematics).
[35] Arutyunov AV. Convexity properties of the Legendre transformation. Math Notes 1980;28(2): 594–595.
[36] Antosik P, Mikusinski J, Sikorski R. Theory of Distributions. The Sequential Approach. Amsterdam: Elsevier Scientific Publishing Co.; 1973.
[37] Krein MG, Rutman MA. Linear operators leaving invariant a cone in a Banach space. Uspehi Matem Nauk (N S) 1948;3(1):3–95.
[38] Vasil'ev FP, Ivanitskii A. Linear Programming. Moscow: Factorial Press; 2003.
[39] Vasil'ev FP. Methods of Optimization. Moscow: Factorial Press; 2002.
[40] Egglston HG. Convexity. Cambridge: Cambridge University Press; 1958. (Cambridge Tracts in Mathematics; vol. 47).
[41] Fedorchuk VV, Filippov VV. General Topology. Basic Constructions. Moscow: Moscow University Press; 1988.
[42] Arutyunov AV, Vartapetov SA, Zhukovskiy SE. Some properties and applications of the Hausdorff distance. J Optim Theory Appl 2016;171(2):527–535.
[43] Fort Jr MK. Points of continuty of semi-continuous functions. Public Mathem, Debrecen 1951;2: 100–102.
[44] Kuratowski K. Topology. New York, London: Academic Press; 1968.
[45] Aseev SM. Approximation of semicontinuous multivalued mappings by continuous ones. Mathematics of the USSR-Izvestiya 1983;20(3):435–448.
[46] Kuratowski K, Ryll-Nardzewski C. A general theorem on selectors. Bull Acad Polon Sci Sèr Sci Math, Astron, Phys 1965;13:397–403.
[47] Castaing C. Sur les multi-applications measurables. Rev Francaise Informat Recherche Opérationnel 1967;1(1):91–126.
[48] Filippov AF. On certain questions in the theory of optimal control. J SIAM Control Ser A 1962;1: 76–84.
[49] Diestel J, Ruess WM, Schachermayer W. Weak compactness in $L_1(\mu, X)$. Proc. Amer. Math. Soc. p. 447–453.
[50] Michael EA. Continuous selections. Ann Math 1956;63(2):361–382.
[51] Michael EA. Dense families of continuous selections. Fund Math 1959;47:173–178.
[52] Michael EA, Pixley C. A unified theorem on continuous selections. Pacific J Math 1980;87(1): 187–188.
[53] Yost D. There can be no Lipschitz version of Michael's selection theorem. Proceedings of the analysis conference, Singapore 1986. (North-Holland Math. Stud.; vol. 150), Amsterdam: North-Holland. p. 295–299.

[54] Arutyunov AV, Zhukovskii SE. Existence and properties of inverse mappings. Proc Steklov Inst Math 2010;271(1):12–22.
[55] Kelley JL. General Topology. New York, Berlin: Springer-Verlag; 1975. (Graduate Texts in Mathematics; vol. 27).
[56] Cellina A. A theorem on the approximation of compact multivalued mappings. Atti Accad Naz Lincei Rend Cl Sci Fis Mat Natur 1969;47(8):429–433.
[57] Cellina A. Approximation of set valued functions and fixed point theorems. Ann Mat Pura Appl 1969;82(4):17–24.
[58] Lasota A, Opial Z. An approximation theorem for multi-valued mappings. Podstawy Sterowania 1971;1:71–75.
[59] Borisovich YG, Gelman BD, Mukhamadiev E, Obukhovskii VV. On the rotation of multi-valued vector fields. Soviet Math Dokl 1969;10:956–958.
[60] Arutyunov AV. Specific selections for set-valued mappings. Dokl Math 2001;63(2):182–184.
[61] Aleksandrov PS, Pasynkov BA. Introduction to Dimension Theory: An Introduction to the Theory of Topological Spaces and the General Theory of Dimension. Moscow: Nauka; 1973.
[62] Filippov AF. Differential Equations with Discontinuous Right-Hand Sides. Moscow: Nauka; 1985.
[63] Blagodatskih VI, Filippov AF. Differential inclusions and optimal control. Trudy Mat Inst AN SSSR 1985;169:194–253.
[64] Arutyunov AV. Covering mappings in metric spaces and fixed points. Dokl Math 2007;76(2): 665–668.
[65] Arutyunov AV. Stability of coincidence points and properties of covering mappings. Math Notes 2009;86(1/2):153–158.
[66] Covitz H, S B Nadler J. Multi-valued contraction mappings in generalized metric spaces. Israel J Math 1970;8:5–11.
[67] Arutyunov A, Avakov E, Gel'man B, Dmitruk A, Obukhovskii V. Locally covering maps in metric spaces and coincidence points. J Fixed Points Theory and Appl 2009;5(1):105–127.
[68] Milnor JW. Topology from the Differentiable Viewpoint. Princeton, NJ: Princeton University Press; 1997. (Princeton Landmarks in Mathematics).
[69] Krasnosel'skii MA, Zabreiko PP. Geometrical Methods of Nonlinear Analysis. Berlin: Springer-Verlag; 1984. (Grundlehren der Mathematischen Wissenschaften; vol. 263).
[70] Krasnosel'skii MA. Topological Methods in the Theory of Nonlinear Integral Equations. New York: The Macmillan Co.; 1964. (A Pergamon Press Book).
[71] Granas A, Dugundji J. Fixed Point Theory. New York: Springer-Verlag; 2003.
[72] Zeidler E. Nonlinear Functional Analysis and its Applications I. Fixed-Point Theorems. New York, Berlin, Heidelberg, Tokyo: Springer-Verlag; 1986.
[73] Bohnenblust HF, Karlin S. On a theorem of Ville. Kuhn HW, Tucker W, editors, Contributions to the Theory of Games. (Ann. of Math. Stud.; vol. 24), Princeton: Princeton Univ. Press; 1950. p. 155–160.
[74] Kakutani S. A generalization of Brouwer's fixed point theorem. Duke Math J 1941;8:457–459.
[75] Hartman P. Ordinary Differential Equations. Boston, MA: Birkhäuser; 1982.

Index

www.ingramcontent.com/pod-product-compliance
Lightning Source LLC
Chambersburg PA
CBHW061421210326
41598CB00035B/6287